T0190213

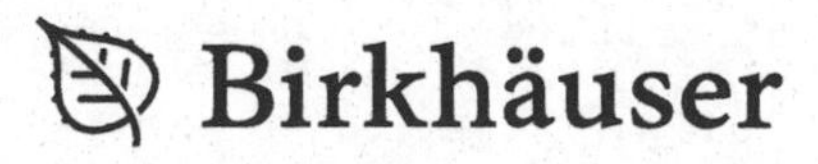
Birkhäuser

Felix Klein
Arnold Sommerfeld

The Theory of the Top
Volume IV

Technical Applications of the Theory of the Top

Raymond J. Nagem
Guido Sandri
Translators

Foreword to Volume IV by Michael Eckert

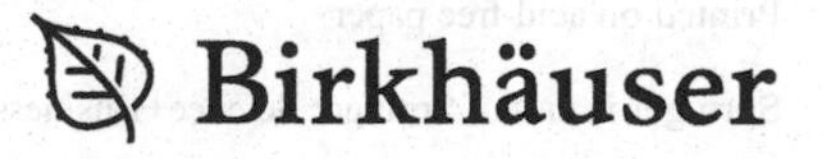

Translators:
Raymond J. Nagem
Department of Mechanical Engineering
Boston University
Boston, MA, USA

Guido Sandri
Department of Mechanical Engineering
Boston University
Boston, MA, USA

ISBN 978-1-4939-5091-1 ISBN 978-0-8176-4829-9 (eBook)
DOI 10.1007/978-0-8176-4829-9
Springer New York Heidelberg Dordrecht London

Mathematics Subject Classification (2010): 01A55, 01A75, 70E05, 70E50, 70E55, 70Q05

Softcover reprint of the hardcover 1st edition 2014

Printed on acid-free paper

Springer is part of Springer Science+Business Media (www.birkhauser-science.com)

Contents

Foreword

Felix Klein recalled in 1922 that *The Theory of the Top* had originated in lectures that were intended to be published as a pamphlet dedicated to the Association for the Advancement of Teaching in Mathematics and the Sciences. "In the hands of my former assistant Sommerfeld, however," Klein wrote, "these lectures were turned into an extensive book" [Klein 1922, p. 509]. The four volumes that were published in 1897, 1898, 1903, and 1910 bear witness to the extent of Sommerfeld's effort, and to the growing departure from Klein's original plan. The more the work considered applications, the less it made use of the mathematical concepts (such as quaternions) introduced in the first volume. Mathematical beauty receded in favor of the complexity involved in addressing astronomical, geophysical, and technical matters. The fourth volume appears almost as an antithesis of the first, with only as much mathematical formalism as necessary in order to cope with one or another technological application.

The time lag between the fourth volume and its predecessors also illustrates the growing divergence from the initial plan. When Klein asked Sommerfeld about the progress of the work in November 1904, one year after the appearance of Volume III, Sommerfeld apologized that he had not had time to think about the top, because he was then busy with research in the theory of electrons. "But the top will still be completed, and for the most part, I think, this winter," he calmed Klein's impatience [Sommerfeld 1904]. Needless to add, this prediction was too optimistic. "I no longer dare to remind you of it," Klein wrote to Sommerfeld again three years later [Klein 1907].

By this time, Sommerfeld was no longer professor at the *Technische Hochschule* in Aachen, where the theory of the top had belonged to the subjects of his own courses. In the summer of 1906, he had accepted a call to the University in Munich as professor of theoretical physics (for Sommerfeld's biography, see [Eckert 2013]). The technical applications of the top intended for the fourth volume, such as the gyrocompass, the stability of the bicycle, and gyroscopic effects in other practical devices, were no longer part of his professorial duties, although he could not resist including such topics in his later Munich lectures on mechanics [Sommerfeld 1952]. However, he had drafted several parts of the manuscript for the fourth volume as early as 1900, when he was called to Aachen, and, in view of a widespread distrust at the *Technische Hochschule* against mathematicians from universities, was eager to comply with the expectations of engineers by choosing gyroscopic applications to ballistics and torpedoes as examples. Furthermore, there were others on whose assistance he could rely, such as Klein's Göttingen colleagues Karl Schwarzschild and Emil Wiechert for the astronomical and geophysical applications in the third volume. For the fourth volume he could count in particular on Fritz Noether, the brother of the famous mathematician Emmy Noether. Fritz Noether had attended Sommerfeld's lectures and considered himself his pupil, although he obtained his doctoral degree in February 1909 in mathematics with Aurel Voss as his thesis advisor and with theoretical physics only as a subsidiary subject [Noether 1909]. In his preface to Volume IV, Sommerfeld thanked Fritz Noether as "an independent collaborator with expertise in mechanics." Noether's assistance was also acknowledged on the title page.

Sommerfeld's and Noether's effort for *The Theory of the Top* did not end with the publication of the fourth volume. One year after its appearance, the publisher informed Sommerfeld that a new edition of the first volume would soon be expedient. The First World War interrupted the swift appearance of new editions, but by 1921 the first three volumes were available in a new printing. There seems to have been no such demand for the fourth volume, because the engineering applications of the theory of the top were subject of a number of more specialized treatises, but this did not reduce Sommerfeld's reputation as an expert in these matters. He was asked, for example, for his expert opinion in patent disputes on gyroscopic compasses [Broelmann 2002, pp. 328–346]. One of his disciples, Karl Glitscher, became employed in the gyrocompass firm of Hermann Anschütz-Kämpfe. In the First World War, Sommerfeld and Noether made use of their expert knowledge in a study of gyroscopic effects in ballistics (described in general terms in the last section of this volume). Fritz Noether made a name for himself as an expert in many fields of applied mathematics. When the Nazis came to power in 1933, he emigrated to the Soviet Union, where he fell victim to Stalin's purges [Schlote 1991]. In August 1940, Sommerfeld asked the Russian Academy for help, but his appeal was to no avail. A year later, Noether was sentenced to death and shot. In his appeal to the Russian Academy, Sommerfeld mentioned that "Noether has provided me with valuable help for the elaboration of the fourth part of the theory of the top edited by Felix Klein and myself," recalling once more Noether's part in this effort [Sommerfeld 1940].

The contributions of Schwarzschild, Wiechert, Noether, and others remind us that voluminous works of this kind are seldom the result of solitary contemplation in the ivory tower of science. If we consider their involvement, *The Theory of the Top* becomes a historical monument for scientists and engineers whose contributions are usually remembered in other areas or forgotten altogether. From the broader perspective of the history of science, *The Theory of the Top* is much more than the legacy of a legendary mathematician and a famous theoretical physicist. Its genesis and the process of its realization demonstrate how technology was regarded by its authors first and foremost as a means for exposing the virtues of science. Although the fourth volume was devoted explicitly to applications in technology, Klein and Sommerfeld still hoped, as they wrote in their 1910 foreword, that the top would prove worthy of the honorable name of a philosophical instrument.

Michael Eckert
Deutsches Museum, Munich

PREFACE.

When F. Klein gave a two-hour lecture "On the Top" in the winter semester of 1895/96, he attempted, in the first place, to emphasize the direct and particularly English conception of mechanical problems, as opposed to the more abstract coloring of the German school, and, on the other hand, to make the particularly German methods of Riemann function theory fruitful in mechanics. The consideration of applications and physical reality would thus be outlined and forcefully advanced in a detailed example, but not yet carried out to full extent.

In the extensive printed edition originating from the pen of A. Sommerfeld, interest in applications prevailed more and more, especially after his appointment to a teaching position in technical mechanics and later in physics. The astronomical, geophysical, and technical content added in this way required, in consequence, the necessity of a change, compared with the original lecture, in the mathematical point of view. While the approximation methods prepared in the first volumes (the method of small oscillations, the treatment of pseudo-regular precession) and the intuitive formulation of the principles of mechanics by means of the impulse concept were perfectly conformable to applications, the advanced function-theoretical methods, the exact representation of the motion by elliptic functions, etc., were later found to be dispensable. Thus, for example, the parameters α, β, γ, δ and their related quaternion quantities, whose geometric meaning was elaborated in the first volume and whose analytic importance was given special emphasis in the second volume, withdrew in the third and fourth volumes, naturally in complete agreement with Klein himself, whose interests had likewise turned more and more toward applications. In particular, the presentation of the technical top problems in the fourth volume used only the very simplest and most elementary law of top motion, which flows immediately from the concept of the impulse in the dynamics of rigid bodies, and which is briefly derived once more at the beginning of this volume.

We would not deny, that with the loss of the unity of time in the course of the fifteen years which have elapsed between the first plan and the present conclusion of the book, our work has also lost its unity of substance and manner of presentation; that what we often promised earlier with respect to the general, so-called analytic mechanics, especially in the advertisements of Vols. I and II, was later not kept; and that we have pursued many mathematical side roads which temporarily diverted us from our primary goal: the concrete understanding of dynamical problems. May the comprehensiveness of the content and the multiplicity of the engaged fields of interest be regarded as substitutes for the lack of systematicness and purposefulness of the presentation.

If we had to dispose of the collected subject material anew, we would probably present the actual mechanics of the top, including its applications, in a much smaller space, by pruning the analytic shoots that branch so joyously from the stem of mechanics. With this presentation we would address the large audience with scientific or technical interests in the theory of the top. The detailed analytic developments, which we would certainly not suppress simply on the basis of their special beauty, would be submitted in another presentation only to the more restricted mathematical circle. As for what pertains, finally, to the requirements of the completely unmathematical reader, and therefore to the difficult question of the popular explanation of the top phenomena, we have taken an extensively grounded critical position in the second volume, and at the beginning of the fourth volume have again pointed out the somewhat long but, it appears to us, only passable way, which begins from the general impulse theorems of rigid-body dynamics. The impulse theorems are either systematically developed from particle mechanics, or, should the occasion arise, illustrated only by experiments, and then postulated axiomatically; on the basis of these theorems, all the partly paradoxical facts of the theory of the top may be understood qualitatively as well-defined approximations, and their domains of validity delimited without want of clarity.

The top is suitable above all other mechanical devices for awakening the sense for true mechanics. May it, in the presentation of our book, serve this purpose in elevated measure, and thus prove worthy in the future of the honorable surname formerly bestowed upon it by Sir John Herschel, the name of a philosophical instrument!

Göttingen and Munich, April 1910.

F. KLEIN. A. SOMMERFELD.

Advertisement to Volume IV of the Theory of the Top.

The original intention of this concluding volume was to depict, in addition to technical applications, the significance of the theory of the top for mechanics and physics in general, along the line of thought due particularly to Maxwell, Lord Kelvin, Helmholtz, and J. J. Thomson.[275] This intention, however, has been renounced, in part because the method of mechanical models does not appear to have the same significance in the current conception of natural science that it had ten years ago, and in part because the analytic development of cyclical systems, generalized coordinates, etc., has already become common scientific knowledge through recent presentations.

Instead, it appeared more useful to devote this concluding volume exclusively to *technical applications of the theory of the top*, applications that all stand in the flow of development and offer great promise for the future. These applications include the gyroscopic effect for high-speed railways, the torpedo guidance apparatus, the gyroscopic compass, the Schlick gyroscopic ship stabilizer, and single-rail trains. Among the further topics of discussion are the stability of the bicycle and gyroscopic effects for turbine engines and ballistics.

The true difficulties of all these problems lie in the technical implementation, the exclusion of perturbing effects, and the control of materials; their theoretical treatment is managed, for the most part, with relatively elementary means. Correspondingly, the character of this part is mathematically much simpler than that of the preceding volumes. The advanced analytic and geometric methods of the first and second volumes became dispensable; the initially chosen concrete form of the fundamental mechanical principles, in contrast, proved particularly valuable here. Its simplicity also makes it possible for the reader to use this concluding part, if necessary, without a detailed study of the preceding volumes.

An extensive consideration of technical applications was obviously essential in order to achieve the original intention of the entire work: to present, in the example of the top, mechanics in its manifold relation to mathematical disciplines on the one hand, and to the natural sciences on the other.

The manuscript of this volume stems in part (§2, §3, and parts of §10) from as early as the year 1900. Because of many other duties, however, I was again and again drawn away from this topic. Even now the work could hardly have been brought to a conclusion if I had not found in Dr. Fritz Noether[276] an independent collaborator with expertise in mechanics. I thank him heartily for his perseverance and enthusiasm. The manuscript of §§5 to 8 and §10, No. 3, originated with him, and was revised only stylistically by me. In particular, the new developments for the ship stabilizer (§5) and the treatment of the bicycle (§8) are his own. The parts of the manuscript originating with me have also been critically edited and supplemented by him. Mr. F. Klein was unfortunately obliged to break off his collaboration for this volume almost completely due to other more pressing demands.

In addition to Mr. Noether and Mr. cand. math. Behrens,[277] I have been assisted in editing by a series of technical specialists. These are the gentlemen[278]

Dr. Anschütz - Kämpfe	for	§7,
Privy councillor Cranz	„	§10, Nos. 7 and 8,
Chief torpedo engineer Diegel	„	§3,
Consul Dr. Schlick	„	§4 to 6,
Dr. Schuler	„	§7,
Professor Skutsch	„	§2 and 6,
Dr. Thoma	„	§10, No. 5.

For their many valuable suggestions, I give my sincere thanks.

Munich, April 1910.

A. Sommerfeld.

Volume IV

Technical Applications of the Theory of the Top

Chapter IX.

Technical applications.

§1. The most important formula in the theory of the top. Generalities on stabilization by the gyroscopic effect.

We cannot regret, in the interest of the subject matter, that this concluding volume of our work has appeared so late. Indeed, the principal technical applications that we consider here have arisen only in the last ten years, during the printing of this book; we recall the high-speed railway, the gyroscopic ship stabilizer, and the gyroscopic compass.

The principle that we previously adopted as the basis of our entire presentation—the preeminence of the concept of the impulse—is particularly confirmed in the explanation of these applications. We arrive immediately from this principle to the formula upon which rests the theory of almost all technical applications.

Since we wish the developments of this chapter to prove fruitful in the hands of engineers and to be understandable without extensive study of the previous chapters, it appears appropriate to summarize here, in brief, the formulas and concepts that serve as their basis.

The impulse (or, more precisely, the "moment of the impulse") is the moment of the quantities of motion of the individual mass elements of the rotating rigid body, with respect to the fixed point of the body. If we consider a rapidly spinning rotor or wheel, then the rotation about the symmetry axis (the "figure axis") so exceeds the other rotations imparted to the body (the "top") that the plane of this moment stands perceptibly perpendicular to the figure axis, and thus falls perceptibly in the "equatorial plane of the top." The impulse vector that is erected perpendicular to this plane then falls perceptibly onto the figure axis. The component of the impulse with respect to this axis, compared to which the other components may generally be neglected, is called the "eigenimpulse" N. The eigenimpulse is the driving moment ("angular momentum") that we transfer to the top when starting it with a string, and which is maintained in the gyroscopic ship stabilizer by the turbine drive or in railway wheels by the pulling force of the locomotive.

Corresponding to the previous preference for left-handed coordinate systems and the clockwise sense of rotations (cf., for example, Fig. 3 on p. 18), we draw the impulse vector toward the side seen from which the rotation about the figure axis, the "eigenrotation," occurs in the clockwise sense.

The introduction of the impulse now provides the possibility of formulating the fundamental laws for the dynamics of the top just as simply, and with almost the same words, as we formulate the fundamental laws for the dynamics of a single mass particle: *the force-free top moves so that its impulse remains constant in magnitude and direction in space* (corresponding to the Galilean law of inertia for a single mass particle); *under the influence of external forces, the top moves in such a way that the rate of change of the impulse vector is equal in magnitude and direction to the moment of the external forces* (likewise represented by a vector) *with respect to the support point of the top* (Newtonian law of acceleration*) for a mass particle).

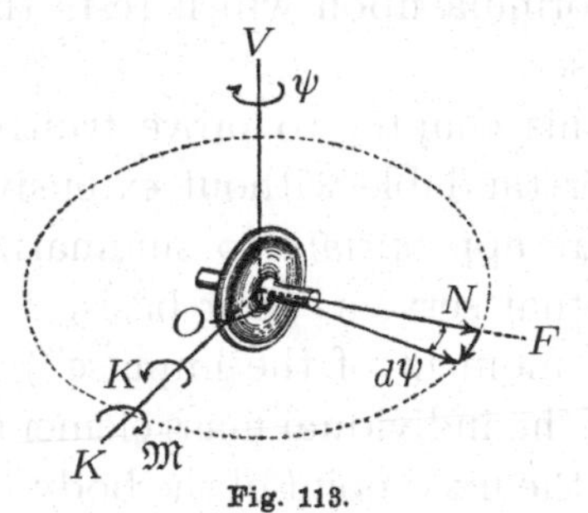

Fig. 113.

I. The most important formula in the theory of the top. We consider a rotor that has been given the eigenimpulse N (equal to the moment of inertia about the figure axis times the angular velocity about the same axis), and imagine that its figure axis is horizontal (Fig. 113). We turn the axis about the vertical through O (turning angle ψ, angular velocity $d\psi/dt$) without influencing the magnitude of the eigenimpulse. What is the required moment?

The answer is provided by our impulse theorem: in the time dt, the endpoint of the vector N describes the path $N\,d\psi$ perpendicular to the figure axis in the horizontal plane. Its rate of change is therefore

$$N\frac{d\psi}{dt}.$$

*) This designation is actually not historically correct. Newton speaks in his lex secunda not of the acceleration, but rather of the change of the "quantitas motus"; that is, the change of the impulse. The concept of the impulse is generally preeminent in the Newtonian principles.

This is, at the same time, the external moment $\mathfrak{M}$ that we must apply in order to effect the change of the impulse. The external moment acts about the axis OK that is perpendicular to the figure axis OF and the vertical OV; we previously called this axis the "line of nodes." The sense of the moment is seen in the figure. If we represent $\mathfrak{M}$ by a vector, then this vector has the same direction as the change of the impulse, and therefore points in the figure to the front. There corresponds a turning arrow that surrounds the semiaxis OK in the clockwise sense.

The moment $\mathfrak{M}$ is opposed by the *gyroscopic effect*; that is, the inertial effect of the spinning rotor that we must continuously overcome if we move the rotor in the described manner. We call this effect K, and have

$$K = N\frac{d\psi}{dt}. \tag{I}$$

The sense of the gyroscopic effect is opposite to that of $\mathfrak{M}$, and is thus counterclockwise.

The gyroscopic effect thus strives to upright the axis of the rotor and place it in equi-orientational parallelism with the axis of the added rotation, in such a manner that the sense of the eigenrotation would coincide with that of the added rotation. The magnitude of this effort is determined by (I); *it is held in equilibrium by the external moment* $\mathfrak{M}$.

It is to be noted that a vertical impulse $A\, d\psi/dt$ is indeed required for the initiation, but not for the maintenance, of the rotation about the vertical ($A =$ the moment of inertia of the top about an equatorial axis). The total impulse of the top thus consists of the resultant of this vertical component and the horizontal eigenimpulse N. In that we neglect the former component and identify the total impulse with the eigenimpulse in the application of our impulse theorem, we commit an error, and implicitly assume that the rotational velocity $d\psi/dt$ is small in comparison with the eigenrotation. Our formula (I) is therefore approximate (cf. the rigorous formula (III)), but is well suited for application to practical cases.

II. Generalization of this formula. We again imagine that the top turns about the vertical, but now in such a way that the figure axis does not sweep the horizontal plane, but rather is inclined to the vertical at an arbitrary angle ϑ. If its length is constant, the eigenimpulse N describes a circular cone, and its endpoint describes a circle of radius $N \sin \vartheta$ (cf. Fig. 114). During the time dt, the path of the impulse endpoint has length $N \sin \vartheta \, d\psi$, and its velocity is $N \sin \vartheta \, d\psi/dt$, parallel to the line of nodes. The magnitude and the axis of the external moment $\mathfrak{M}$ that is to be applied for the maintenance of the motion are thus determined, as are the magnitude and axis of the *gyroscopic effect*

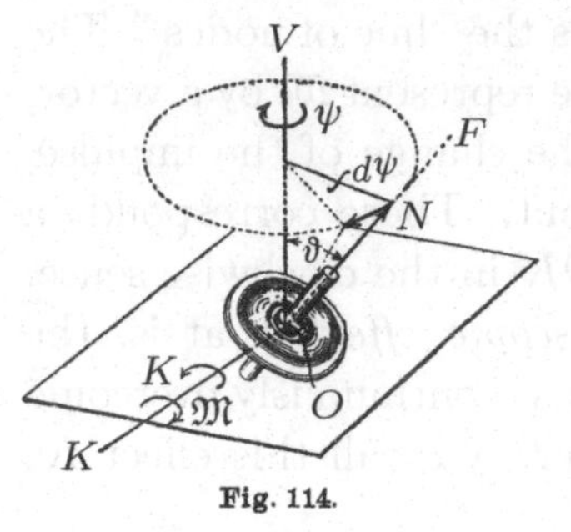

Fig. 114.

$$K = N \sin \vartheta \frac{d\psi}{dt} \tag{II}$$

with which the top resists the rotation about the vertical. The sense of the gyroscopic effect is again (cf. Fig. 114) described by the tendency to equi-orientational parallelism. Formula (II) is only approximately valid; it assumes that the eigenrotation is large compared with the added rotation.

III. Rigorous expression for the inertial effect of the top. In spite of its limited importance, the general and rigorously valid value of the inertial effect, which is composed of the previously considered gyroscopic effect and an additional centrifugal effect, is also derived under the assumptions that the figure axis is inclined by the angle ϑ with respect to the vertical and the rotation $d\psi/dt$ is uniform, so that the impulse moves without change of its magnitude on a circular cone about the vertical.

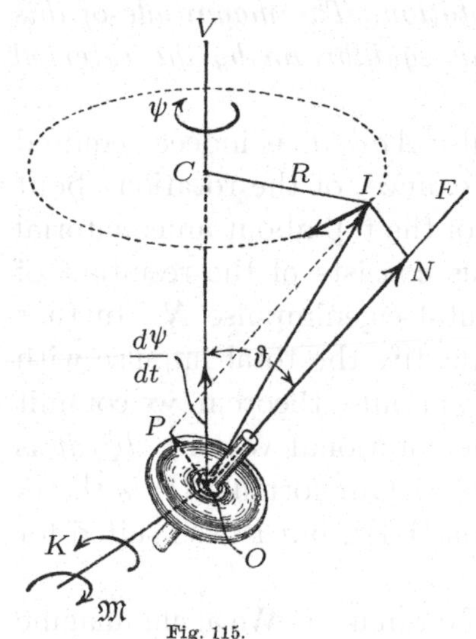

Fig. 115.

The path element of the impulse endpoint I (cf. Fig. 115) is $R \, d\psi$, where $R = CI$ is the radius of the circle described by I. The applied moment $\mathfrak{M}$ and the opposed inertial effect K are determined as above. They therefore depend merely on the calculation of the radius R.

We draw the angular velocity $d\psi/dt$ as a rotation arrow in the vertical direction and decompose this arrow, since the vertical is not a principal axis for the mass distribution of the rotor, into the two components

$$\sin \vartheta \frac{d\psi}{dt} \quad \text{and} \quad \cos \vartheta \frac{d\psi}{dt}$$

with respect to the equatorial plane and the figure axis. The corresponding components of the impulse are obtained by multiplication with the respective moments of inertia. If the moment of inertia about an equatorial axis is denoted by A, then the component of the impulse perpendicular to the figure axis, which was neglected with respect to the eigenimpulse in the preceding approximate consideration, is

$$OP = A \sin\vartheta \frac{d\psi}{dt}.$$

The endpoint of the impulse vector I is obtained if one sets $OP = NI$ in the figure and adds this component that is perpendicular to the figure axis to the eigenimpulse $N = ON$. The radius R is now given, if we project the line segments ONI onto the direction of CI, by

$$R = ON \sin\vartheta - NI \cos\vartheta = N \sin\vartheta - A \sin\vartheta \cos\vartheta \frac{d\psi}{dt}.$$

The desired inertial effect is thus

$$\text{(III)} \qquad K = R\frac{d\psi}{dt} = \left(N - A\cos\vartheta \frac{d\psi}{dt}\right) \sin\vartheta \frac{d\psi}{dt}.$$

The additional term

$$-A \sin\vartheta \cos\vartheta \left(\frac{d\psi}{dt}\right)^2$$

that is found here is known from the theory of the simple spherical pendulum; it is designated there as the moment of the centrifugal force. If, namely, the eigenimpulse of the top vanishes ($N = 0$), then the top swings like a spherical pendulum with moment of inertia A. We can imagine realizing this by a simple pendulum with length l and mass m, so that

$$ml^2 = A.$$

The horizontal centrifugal force is then

$$Z = ml \sin\vartheta \left(\frac{d\psi}{dt}\right)^2,$$

and its moment about the line of nodes is

$$ml \sin\vartheta \left(\frac{d\psi}{dt}\right)^2 l \cos\vartheta = A \sin\vartheta \cos\vartheta \left(\frac{d\psi}{dt}\right)^2,$$

which is exactly the expression above. The negative sign of our additional term also conforms with this consideration, in that the moment of the centrifugal force of the pendulum strives to distance it from the vertical, and thus has the opposite sense as the first term in (III) and the arrow K in Fig. 115.

It is to be remarked, however, that the separation of the inertial effect (III) into the gyroscopic effect and the centrifugal effect is not absolute, but rather depends on the fact that we have distinguished the figure axis in the computation of the impulse.

IV. Stabilization by the gyroscopic effect. One of the most striking and well-known consequences of the theory of the top is the possibility of stabilizing an unstable or neutral degree of freedom by the installation of a rotating mass. The schema of this procedure, for which the discussions of this chapter will provide many examples, may be represented on the basis of our formula (I) by an example in the following manner.

The position of the figure axis in the horizontal plane (cf. Fig. 116), which is measured by the angle ψ, is, in itself (that is, for a nonspinning rotor), indifferent: a turning-moment Ψ about the vertical causes an angular displacement ψ that is determined in terms of the equatorial moment of inertia A about the vertical by the acceleration equation

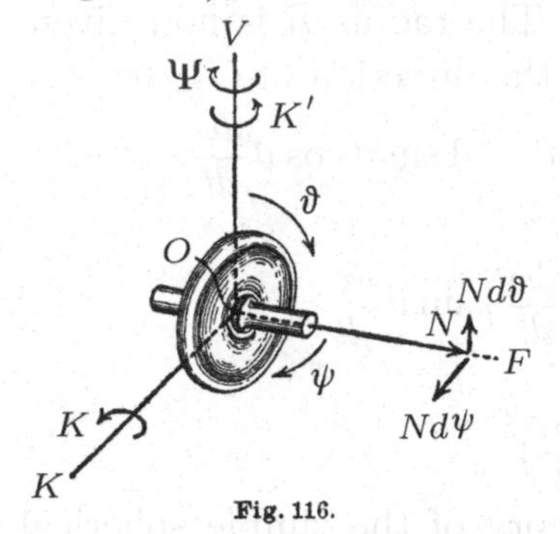

Fig. 116.

$$\text{(IVa)} \qquad A\frac{d^2\psi}{dt^2} = \Psi.$$

If the rotor spins, however, then this angular displacement is bound with a gyroscopic effect K that is determined by (I). *We assume that the figure axis is no longer fixed in the horizontal plane*, as we assumed in (I), *but rather that it can follow, by means of an appropriate suspension, its tendency to parallelism with the vertical rotation axis.* The inclination ϑ with respect to the vertical will then decrease according to the acceleration equation

$$\text{(IVb)} \qquad A\frac{d^2\vartheta}{dt^2} = -K = -N\frac{d\psi}{dt}.$$

The angular velocity of the elevation of the figure axis is thus

$$\text{(IVc)} \qquad \frac{d\vartheta}{dt} = -\frac{N}{A}\psi,$$

assuming that we measure the azimuth ψ from a position in which $d\vartheta/dt$ is equal to zero.

Now, however, the continuation of the rotation $d\vartheta/dt$ about OK would require an upward change of position of the impulse N, just as the rotation $d\psi/dt$ requires a change of position of N in the horizontal plane. Thus a gyroscopic effect again occurs, and indeed about the axis OV that is perpendicular to OK and OF. We call this gyroscopic effect K', and again determine it by formula (I) as

$$\text{(IVd)} \qquad K' = N\frac{d\vartheta}{dt}.$$

Corresponding to the tendency to equi-orientational parallelism, the gyroscopic effect K' acts in the opposite sense to that of the original rotation $\frac{d\psi}{dt}$, and is therefore counterclockwise in Fig. 116. The countermoment K' is therefore joined to the turning-moment Ψ.

The equation of motion (IVa) now becomes[279]

$$\text{(IVe)} \qquad A\frac{d^2\psi}{dt^2} = \Psi + K' = \Psi + N\frac{d\vartheta}{dt}.$$

The schema of stabilization by the gyroscopic effect is now given in equations (IVe) and (IVb). With the use of equation (IVc), namely, equation (IVe) becomes

$$\text{(IV)} \qquad A \cdot \frac{d^2\psi}{dt^2} = \Psi - \frac{N^2}{A}\psi.$$

Equation (IV) expresses the possibility of a stable oscillation in the coordinate ψ.

If, namely, the eigenimpulse N is sufficiently large, then the countermoment K' exceeds the original deflecting moment Ψ to such an extent that we can, for qualitative questions, neglect Ψ with respect to K'. The right-hand side of (IV) is then negative for positive ψ, and positive for negative ψ.

The axis of the rotor therefore strives to return to its initial position, exactly like a stable pendulum; it overruns this position, again approaches it, etc. It possesses "a specific capability for resistance against a change of direction, a kind of absolute orientation in space," as stated at the conclusion of Chapter III.

Our derivation of this possibility of stabilization reveals, however, an additional necessary condition of fundamental significance. The axis of the top must have the possibility to deflect in the vertical direction; the rotation $d\vartheta/dt$ must actually be made possible by the suspension of the rotor. If the axis of the top remains restricted to the horizontal plane, then $d\vartheta/dt$ and our stabilizing countermoment K' become zero, and only the gyroscopic effect K occurs, which is expressed as a pressure against the guide that restricts the freedom of motion of the top. The top must therefore have full freedom of motion (two degrees of freedom for the motion of the figure axis in the horizontal and vertical directions, and a third degree of freedom for rotation about the figure axis) if it is to act as a stabilizer. We thus state the following principle:

Stabilization is possible only by a top with three degrees of freedom. If one degree of freedom of the top is eliminated, then the gyro-

scopic effect produced by this degree of freedom ceases to act, and the "specific capability of resistance against a change of direction" is lost. A top with t w o degrees of freedom thus follows the impulses that act on it without resistance. If one degree of freedom is impeded but not entirely eliminated, there remains a certain capability of resistance that is smaller than that for unimpeded freedom of motion.

If, for example, we clamp the inner ring in Fig. 23 and thus eliminate the possibility of rotation about the axis ST, the handle may be turned, when the rotor is spinning, as if it did not contain a top. The reaction forces on the ring then produce the change of position of the impulse vector that is required by the guidance of the figure axis on the prescribed path. If, on the other hand, the axis ST rotates in its bearings with considerable friction, then the countermoment K' will be significantly smaller than what corresponds to equation (IV). In this case, namely, the available impulse change that is given by the right-hand side of equation (IVd) is applied only in small part for the generation of the angular velocity $d\psi/dt$, and in large part for the overcoming of the friction that opposes the rotation $d\vartheta/dt$.

Equations (IV), since they are derived on the basis of formula (I), are burdened with an imprecision to which we will immediately return, and hold only for the time interval in which the figure axis stands approximately perpendicular to the vertical. For now, a qualitative conclusion may be drawn with the use of our imprecise equations.

If we assume, for example, that the deflecting moment Ψ is constant, then equation (IV) can be integrated immediately, and yields

$$\psi = \frac{A\Psi}{N^2} + a\cos\frac{N}{A}t + b\sin\frac{N}{A}t;$$

if, moreover, we set ψ and $\dfrac{d\psi}{dt}$ equal to zero for $t = 0$, then $b = 0$ and $a = -A\Psi/N^2$, and therefore

$$\text{(IVf)} \qquad \psi = \frac{A\Psi}{N^2}\Big(1 - \cos\frac{N}{A}t\Big).$$

The figure axis thus yields somewhat to the moment Ψ, and indeed in the mean, for large N, by the small angle $\psi_m = A\Psi/N^2$; the turning angle oscillates periodically between its original value 0 and the maximum value $2\psi_m$.

At the same time, however, the figure axis is elevated; according to equation (IVc), namely,

$$\frac{d\vartheta}{dt} = -\frac{\Psi}{N}\Big(1 - \cos\frac{N}{A}t\Big),$$

so that if ϑ is initially equal to $\pi/2$, then

$$\text{(IVg)} \qquad \vartheta = \frac{\pi}{2} - \frac{\Psi}{N}t + \frac{A\Psi}{N^2}\sin\frac{N}{A}t.$$

This motion is not purely periodic, but rather has a principal component given by the first two terms that increases with time and is overlaid with small oscillations represented by the last term. The amplitude and period of this oscillation coincide with those of the angle ψ.

For very large N, both oscillations become imperceptibly small (in the same manner as the nutation of the pseudoregular precession in Ch. V, §2), and the elevation of the axis of the top occurs very slowly. Only in this case are the assumptions of our calculation for the application of the formula (I) fulfilled with sufficient accuracy for a not too long observation time t.

If, on the other hand, we would proceed rigorously and consider the elevation of the figure axis that appears in our last example, we must determine the gyroscopic effect K not from equation (I), but rather from equation (III); further, we must take from the gyroscopic effect K', which has as its axis the common perpendicular to OF and OK, only the component that acts about OV. Then, however, the eigenimpulse N is also not constant, but rather will be changed by the turning-moment Ψ about the vertical as soon as the figure axis is no longer perpendicular to the vertical. Finally, the vertical is no longer a principal axis of the mass distribution as soon as the equatorial plane of the top no longer passes through the vertical; thus the acceleration effect of the turning-moment Ψ is no longer given by $A\,d^2\psi/dt^2$, as in equation (IVa), but rather must be determined according to the general impulse theorem and the general relation between the impulse vector and the rotation vector.

These exact equations of motion can indeed be written according to the schema of the Lagrange equations, but they no longer permit of further integration. Nevertheless, the general character of the phenomena can only be, without doubt, the following.

As the angle ϑ diminishes from $\pi/2$ to 0, the gyroscopic effect K and the relevant component of K' successively decrease. The stability of the top with respect to the external moment Ψ is correspondingly reduced, until it vanishes completely for $\vartheta = 0$. In this latter limiting case, where the eigenimpulse is vertical, a change of position of N and a a gyroscopic effect no longer occur. Experience with every model of the top completely confirms this conclusion:

For a progressive elevation of the figure axis, there is a progressive diminishment of stability with respect to a moment Ψ about the vertical.

The following references may serve to relate the preceding with the developments of the previous volumes.

The fundamental impulse theorems are established in Ch. II, §5.

To I. Formula (I) is discussed, for example, in the review of the popular top literature on p. 311, and is subsumed under the general concept of the "deviation resistance for regular precession" in Ch. III, §6. In fact, the motion considered in (I) is a regular precession about the vertical with the inclination angle $\vartheta = \pi/2$, and formula (I) is identical with equation (1) of p. 175 under this condition and within the limit of precision stated above. Concerning the rule of equi-orientational parallelism, see Ch. VIII, p. 734.

To III. Expression (III) coincides with equation (1) of p. 175 for the deviation resistance not only approximately for large N, but rather exactly, in that $N = C(\mu + \nu\cos\vartheta)$; we have presently suppressed the negative sign in that equation, since we now prefer to establish the sense of the gyroscopic effect by the rule of equi-orientational parallelism in a manner that is valid for all cases. Our formula (III) is also contained in the Lagrange equations (p. 154, equations (1)). Our gyroscopic effect K is indeed oppositely equal, according to its definition, to the external moment Θ that acts about the line of nodes, the moment that is required for the maintenance of the regular precession. The ϑ-component of the Lagrange equations for regular precession (ϑ = const.) thus becomes

$$-K = \frac{d[\Theta]}{dt} - \frac{\partial T}{\partial \vartheta}.$$

For ϑ = const., however, expression (6) on p. 156 gives

$$[\Theta] = \frac{\partial T}{\partial \vartheta'} = A\vartheta' = 0,$$

$$\frac{\partial T}{\partial \vartheta} = A\sin\vartheta\cos\vartheta\,\psi'^2 - C(\varphi' + \cos\vartheta\,\psi')\sin\vartheta\,\psi' = (A\cos\vartheta\,\psi' - N)\sin\vartheta\,\psi',$$

and thus

$$-K = (N - A\cos\vartheta\,\psi')\sin\vartheta\,\psi',$$

in agreement with equation (III) (up to the sign not written there).

In the general case that the motion is not a regular precession, and thus the external moment does not exactly maintain equilibrium with the gyroscopic effect, we have

$$[\Theta] = A\vartheta'; \qquad \frac{d[\Theta]}{dt} = A\vartheta'',$$

and thus the Lagrange equation for the ϑ-component is

$$A\vartheta'' = (A\cos\vartheta\,\psi' - N)\sin\vartheta\,\psi' + \Theta,$$

$$\text{or}\quad A\vartheta'' = -K + \Theta,$$

in agreement with equation (IVb) (in which the external moment Θ was assumed to vanish).

To IV. Our present expression for the countermoment of the gyroscopic effect in (IV) naturally agrees, apart from the notation, with the previous equation (3) of p. 195.

Lord K e l v i n's "gyrostats" are realizations of devices in which an unstable degree of freedom is stabilized by gyroscopic effects. A simple example of a gyrostat is depicted in "Natural Philosophy," Vol. I, Art. 345, 2nd ed., p. 397, as well as in the Enc. d. math. Wiss., Bd. IV, Art. 6 (Stäckel), Nr. 43b, p. 675.[280] The above theory of gyroscopic stabilization also gives, with a small modification (the external moment acts on the ϑ-coordinate instead of the ψ-coordinate, and thus is not fixed in space), the theory of that simple gyrostat. The gyrostat, however, is essentially different from our example with respect to the duration of the stabilization; cf. §10, No. 3 of this chapter.

The principle of two and three degrees of freedom emphasized above is stated generally by Lord K e l v i n for an arbitrary number of degrees of freedom, and indeed in the context of the treatment of his gyrostat. It is shown that *only an even number of unstable degrees of freedom* can be stabilized by the inertial effects of cyclic motions, which one may imagine as generalized gyroscopic effects realized by systems of revolving rotors; the number of previously and still remaining stable degrees of freedom, in contrast, is arbitrary. These effects appear analytically in the occurrence of "gyroscopic terms," which are terms in the equations of motion of the noncyclic coordinates that are composed of the noncyclic velocities and the cyclic impulses. See Natural Philosophy, Art. 345^{VI} ff. The simplest examples of these terms are the gyroscopic effects designated in the text by K and K'. Concerning the more precise analytic manner of formation of these terms, cf. the conclusion of §4. An interesting application of Kelvin's theorem in the discussion of the monorail (§10, No. 3) will make the meaning of these terms more clear.

A thorough mathematical investigation of the stabilization problem on the basis of the rigorous L a g r a n g e equations would be very deserving of thanks. The question is simply this: how does the top behave if a (perhaps constant) moment about a *spatially fixed* axis acts on it? The behavior of the top under the application of a moment about an axis that is *fixed in the top* is easily treated by means of the E u l e r equations (cf. the analogous problem in the previous chapter on pp. 728 and 726). The classical problem of the heavy top treats of a moment that is neither *fixed in space* nor *fixed in the top*, but rather stands perpendicular to the spatially fixed vertical and the figure axis that is fixed in the top. In the posing of the problem—but unfortunately not in the mathematical execution—the presently considered stabilization problem is at least as simple and appealing as the classical problem of the top.[281]

§2. Gyroscopic effects in railway operation.

Among the technical applications of the theory of the top, perhaps the simplest concern the effects of rapidly rotating wheels on vehicles.

We consider a railway train, and direct our attention to a single axle. If M is the mass of the carriage (or locomotive) that is carried by our axle, its own mass included, then Mg is the total force with which the wheel-set is pressed against the rail on a horizontal section of the track. The moment of inertia of the wheel-set about its centerline is called C; we understand by r the radius of the wheel, calculated from the centerline to the running surface (the "tread radius"), and set

$$C = mr^2; \tag{1}$$

m is then called the mass of the wheel-set reduced to the circumference of the wheels. If v is the speed of travel of the train, then the angular velocity of the wheels is v/r, and their impulse is

$$N = C\frac{v}{r} = mrv. \tag{2}$$

We can designate the wheel-set as a "symmetric top"; the centerline of the wheel-set corresponds to the figure axis of the top. Gyroscopic effects arise as soon as the direction of the centerline is changed in any way.

a) We first consider the case in which a change of direction is produced by the curvature of the track. We first disregard, for simplicity, the usual superelevation of the outer rail of the curve. Let R be the radius of curvature of the track. The angular velocity in the curve is called $d\psi/dt$, and is calculated from R and the speed of travel v as

$$\frac{d\psi}{dt} = \frac{v}{R}. \tag{3}$$

The wheel-set undergoes two simultaneous rotations in the curve, the "eigenrotation" about its centerline and the "added" rotation $\frac{d\psi}{dt}$ about the vertical axis through the instantaneous center of curvature of the curve. Since we can disregard the forward motion of the train in the calculation of the gyroscopic effect and need only consider the rotation of the wheel-set about its center of gravity, we prefer to assign the axis of the added rotation as the vertical line through the center of gravity of the wheel-set.

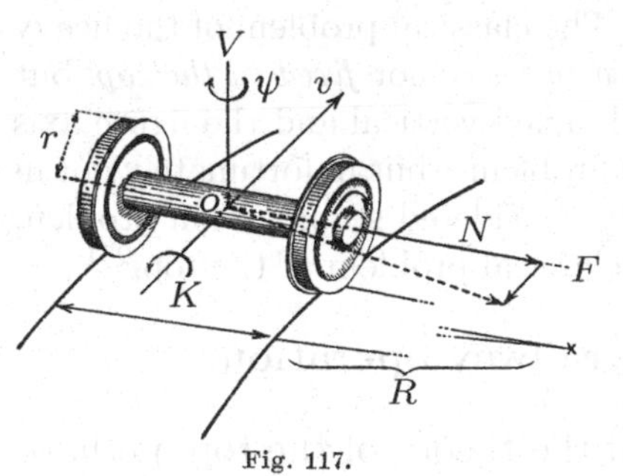

Fig. 117.

The axes OV (the axis of the added rotation $\frac{d\psi}{dt}$) and OF (the axis of the eigenrotation = the figure axis of the top = the centerline of the wheel-set) are marked in Fig. 117. The direction of travel of the train is perpendicular to both, and is imagined in the figure to point away from the viewer, so that the eigenrotation about the rightward axis OF

occurs in the clockwise sense. The midpoint of the curve lies to the right of the direction of travel, so that the rotation $d\psi/dt$ occurs in the clockwise sense as seen from above. We mark in the figure the axis OK, the "line of nodes," which is perpendicular to OV and OF and which coincides, disregarding the sense, with the direction of travel. This is, at the same time, the axis of the gyroscopic effect, which, according to the rule of equi-orientational parallelism (cf. the preceding section), tends to tilt the wheel-set toward the rail that is farthest from the center of curvature, the *outer* (*left-hand* in the figure) rail. This gyroscopic effect is the embodiment of the inertial effects that come into play here for the rotating wheel-set. Its magnitude is determined by formula (I) of the previous section as

$$K = N\frac{d\psi}{dt},$$

or, using equations (2) and (3) of this section,*)

$$K = mv^2\frac{r}{R}. \tag{4}$$

If we denote the gauge of the rails by s and set $K = Ps$, then P signifies the amount that the outer rail is loaded and the inner rail is unloaded by the gyroscopic effect (cf. Fig. 118). The pair of forces P will naturally be taken up by the rails, in the sense that the reaction against the outer rail will be increased by the amount $Q = P$, and the reaction against the inner rail will be diminished by the same amount—both with respect to the reaction force caused by the weight of the vehicle.

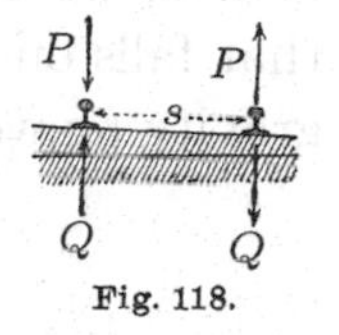

Fig. 118.

The force-pair Qs of the rail reactions is nothing other than the moment $\mathfrak{M}$ that was discussed in the previous section. Its magnitude is equal, according to the general impulse theorems, to the rate of change of the impulse (cf. Fig. 117, where this change is indicated). While the gyroscopic effect K is a fictitious or inertial effect, we must regard the moment $\mathfrak{M}$ of the rail reactions as the physically existing effect by which the guidance of the wheel-set along the rails is realized.

For what concerns the inaccuracy of the formula for the gyroscopic effect that was emphasized in the previous section, we can easily convince ourselves that this inaccuracy is unimportant for the conditions

*) We find this formula, for example, in the stimulating little book of Worthington: Dynamics of rotation, an elementary introduction to rigid dynamics, 5th ed., London 1904, p. 157, Example (1).[282]

of railway operation. The inaccuracy consists in neglecting the impulse of the added rotation in comparison with the eigenimpulse. Now the angular velocity v/R of the added rotation is to the angular velocity v/r of the eigenrotation as the radius of the wheel is to the radius of the curve; the ratio of the corresponding impulse components differs from this ratio only by the factor of the ratio of the moment of inertia about the vertical to that about the figure axis. The validity of the relevant omission is therefore beyond doubt.

There now arises the question of whether our gyroscopic effect is of any practical importance. We best inform ourselves of this if we compare it with the *moment of the centrifugal force* that occurs in the passage through the curve, and which is known to play an important role in railway operation.

We first see that the gyroscopic effect coincides in sense with the centrifugal effect; the centrifugal effect also tends to tip the carriage toward the outer rail. *Because of the gyroscopic effect, the outer rail is still more loaded and the inner rail still less loaded than they are due to the centrifugal effect alone.* It is the latter circumstance, in particular, that is undesirable on the grounds of safety.

The magnitude of the centrifugal force is, in our notation, Mv^2/R; it acts, we can say, at the center of gravity of the part of the vehicle that falls on our axle, and produces, if h is the height of this center of gravity above the surface of the rail, the tipping moment

$$H = Mv^2 \frac{h}{R}.$$

This moment is therefore similar to the moment of the gyroscopic effect not only in its direction, but also in its dependence on the speed of travel and the curvature of the path.

With consideration of equation (4), there follows*)

$$\frac{K}{H} = \frac{m}{M}\frac{r}{h}. \tag{5}$$

Both factors on the right-hand side are proper fractions. For the *total mass* of the considered portion of the vehicle is greater than the *actual* mass of the wheel-set, and this actual mass is again greater than

*) Cf. F. Kötter: *Die Kreiselwirkung der Räderpaare bei regelmäſsiger Bewegung des Wagens in kreisförmigen Bahnen.* Sitzungsber. d. Berliner Mathem. Gesellschaft, III. Jahrgang, 1904, p. 36, where a uniform elevation of the outer rail is also considered. Mr. Kötter interprets our ratio (5) of the gyroscopic effect to the centrifugal effect as a proportional increase of the lever arm of the centrifugal effect; namely, as an apparent elevation of the center of mass of the carriage.[283]

the *reduced* mass of the wheel-set (that is, the mass assigned to the circumference of the wheels that gives the same moment of inertia about the centerline as the actual mass of the wheel-set). Further, the center of gravity of the wheel-set lies at the height r above the rails, and therefore the height of the overall center of mass of the considered potion of the vehicle is always greater than r. Thus K is certainly less than H.

The gyroscopic effect will be relatively largest for electrically driven trains, since here, as in the following example, the motor is fixed directly to the wheel-set, and thus the moment of inertia and the reduced mass of the wheel-set will be relatively large. Moreover, the center of gravity of the total mass is relatively low, so that the centrifugal effect will be small. The following data correspond to a high-speed rail carriage built by S i e m e n s and H a l s k e that was used for the trial runs of the *Studiengesellschaft für elektrische Schnellbahnen* on the Marienfelde-Zossen line in August 1903.*) The weight on an individual axle was about 15 tons,**) the actual weight of a driving wheel-set was 4 tons and its weight reduced to the circumference about 1,5 tons, the height of the center of mass above the rail surface***) was about 1 m, and the wheel radius was 0,625 m.

We thus have $m/M = 1{,}5/15 = 1/10$, $r/h = 0{,}625$; there follows

$$\frac{K}{H} = 0{,}0625 = \frac{1}{16}.$$

The total moment in the passage through a curve is therefore

$$H + K = H\left(1 + \frac{1}{16}\right) = 1{,}06\, H,$$

and is thus, under the assumed conditions, about 6% higher than the usually considered moment of the centrifugal force alone. This total moment increases, because of the expression for H, in a well-known manner with the curvature and, in particular, with the speed of travel, while the ratio of the two components K and H is independent of the curvature and the speed of travel.

*) Cf. Elektrotechn. Zeitschr. 1901, p. 778, or Zeitschr. d. Ver. deutsche Ing. 1904, p. 949. The motor is not fixed to the axle in more recent designs; one now seeks the best possible spring suspension of the motor.

**) The total weight of the carriage was 93 400 kg; the carriage had two three-axle bogies, each with two outer motor axes and an undriven axis in the middle.

***) According to an obliging communication from the manufacturing firm.

b) It is well known that the detrimental moment of the centrifugal force may be counteracted by the superelevation of the outer rail in a curve, thus bringing into play an opposing gravitational moment. Because of the previously considered gyroscopic effect, this superelevation, as we saw above, must be increased by a few per cent.

The necessary superelevation now has, however, a further consequence. The superelevation must naturally begin and end gradually, in so-called transition curves. The inclination of the axle of the wheel-set thus continuously increases or decreases as it enters or leaves the curve. Evidently, a gyroscopic effect with a *vertical* axis (different from that previously considered) now appears.

We begin from Fig. 119, and imagine, just as in Fig. 117, that the center of curvature of the track lies to the right, so that the left (outer) rail is superelevated. In the passage from the nonsuperelevated to the superelevated section, the wheel-set will rotate about the axis of the direction of travel in the clockwise sense; the corresponding angular velocity is $\frac{d\varphi}{dt}$. Because of this angular velocity, the impulse N of the wheel-set will be deflected downward; the required turning-moment is, judged from the half-line OV, negative. The opposite turning-moment represents the resistance of the wheel-set against the deflection of the impulse, or, concisely, the *gyroscopic effect*; this effect therefore acts, corresponding to the rule of homologous parallelism, about OV in the clockwise sense. This is indicated by the attached arrow K' in Fig. 119. If the transition curve runs, in contrast, from a superelevated curve to a nonsuperelevated straight section, the sense of $\frac{d\varphi}{dt}$ is evidently reversed, and thus also the sense of the gyroscopic effect.

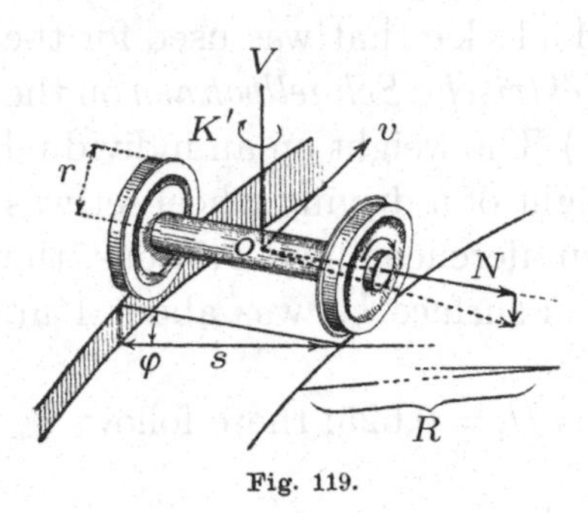

Fig. 119.

The magnitude of our present gyroscopic effect is again determined by formula (I) of the previous section as

$$K' = N\frac{d\varphi}{dt}. \tag{6}$$

In order to compare the magnitude of K' with the previously calculated magnitude of K, we must first calculate the angular velocity $\frac{d\varphi}{dt}$. Let h' be the superelevation of the outer rail in the circular curve and l be the length of the transition curve; h'/l is then the slope, assumed to be uniform for the transition curve, of the outer rail with respect to the horizontal plane, or also the ascent of the outer rail per unit length. If s is again the gauge of the track, then the change in angle of the centerline of the wheel-set, likewise per unit length, is h'/ls. Since the vehicle covers the distance v per unit time, the change in angle per unit time (that is, the angular velocity $\frac{d\varphi}{dt}$) is

$$\frac{h'v}{ls}.$$

If we substitute for N the value from equation (2), then (6) gives

$$K' = mv^2 \frac{h'}{l} \frac{r}{s}. \tag{7}$$

The moment K', just like K, naturally increases with the speed of travel, and depends, moreover, on the greater or lesser slope of the rail in the transition curve. The comparison of formulas (4) and (7) now shows that

$$\frac{K'}{K} = \frac{h'}{l} \frac{R}{s}. \tag{8}$$

In the passage from a straight to a circular section of the track, the radius of curvature R of the transition curve will be diminished continuously from its initial value ∞ to the radius of the circular section. Since the gyroscopic effect K increases with decreasing radius of curvature and the gyroscopic effect K' is independent of the radius of curvature, one sees that K' must predominate at the beginning of the transition curve (R very large). We ask whether this is also true at the end of the transition curve where K has reached its full value, which is decisive for the further passage of the circular curve.

We assume $R = 900$ m, the sharpest curvature that may be passed in railway operation at full express speed (100 km/hour). The gauge is s = 1,435 m throughout. In the layout of the track, the length of the transition curve is chosen so that the slope is not more than 1 : 300; that is, one chooses l, as far as possible, equal to $300h'$. Formula (4) thus gives

$$\frac{K'}{K} = \frac{900}{430} = 2{,}1.$$

It is thus seen that the gyroscopic effect about the vertical is always greater than that about the direction of travel in the passage through a transition curve. When the circular part of the curve is reached, in which the change of the superelevation no longer occurs, the effect K' naturally ceases, and K alone remains.

In the report*) of the previously mentioned *Studiengesellschaft für elektrische Schnellbahnen* on their trial runs of September–November 1903, we find, in fact, an indication of the importance of gently inclined transition curves for high-speed travel. The radius of curvature of the sharpest curve on the test track was 2000 m; the 80 mm superelevation of the outer rail was originally graded on a section of length 50 m; that is, with a slope of $1:600$. This slope, however, proved too steep for speeds greater than 160 km/hr; in the entrance to the curve, the one-sided elevation of the carriage was clearly perceived as a jolt whose turning component could indeed be attributed primarily to the gyroscopic effect. When the slope was reduced to $1:1200$, in contrast, the observed jolt vanished almost completely.

It therefore appears that in contrast to the originally considered gyroscopic effect K, whose immediate practical significance is small, the gyroscopic effect K' can have undesirable consequences. We thus wish to pursue the type and manner of formation of these consequences in more detail.

It is well known that the wheel-set always has some play (10 to 20 mm) between the rails; it can therefore respond to the gyroscopic effect K', which tends to turn it about the vertical, within restricted limits, while, however, it is opposed by very strong friction between the wheels and the rails. It may be noted that the friction is essentially independent of velocity, while the gyroscopic effect increases quadratically with velocity. As the velocity increases , the gyroscopic effect K' could therefore be expressed, in spite of the large friction, as a rotation about the vertical. A similar rotation about the vertical, or, as one says in railway operations, a *lurching*, will then be produced in the vehicle that is attached to the wheel-set. Because of the elastic coupling between the wheel-set and the vehicle, such a lurching motion can, especially for a locomotive, persist for a few oscillations once it is established, and thus, in its turn, act in reverse on the wheel-set to an extent that is naturally again determined by the rail friction. In the same manner as the originally considered travel through a curve, a deflection of the wheel-set about the vertical produces a gyroscopic effect K, which consists of a moment about the direction of travel.

*) Berlin 1904 by S. Hermann, p. 44.[284]

The magnitude of the resulting moment depends on the velocity of the lurching. The latter, however, will now be significantly larger than for traveling through a curve, where it amounted to only v/R. It may be emphasized that the angular velocity of the lurching may be reduced if one gives the vehicle the greatest possible moment of inertia about the vertical.

c) In our presentation, the intentional superelevation of one rail in a transition curve was regarded as the primary cause of the moment K'. It is evident, however, that an unintentional difference in the height of the rails and a nonuniformity in the circumference of the wheels will act in the same sense, and perhaps to a greater degree. If, for example, one wheel on our axle is *out-of-round*, then the centerline of the axle tilts slightly upward and downward during each rotation of the wheel. The resulting gyroscopic effect K' depends not on the magnitude of the descent or ascent itself, but rather on the magnitude of the falling and rising: this magnitude will be relatively large for small irregularities in the circular form of the wheel, since the rise and fall are repeated on each relatively short section of track that corresponds to the circumference of the wheel. Strong moments K' and rapid lurching oscillations can occur in this manner, finally resulting in a significant moment K and a considerable alternating unloading of one or the other rail. *Nonuniformity in the vertical rail guidance*, or so-called track heaving, acts in the same manner as irregularity in the circumference of the wheel. *Nonuniformity in the horizontal rail guidance*, or lateral buckling of the rails, causes the moment K in a more direct manner, and thus causes the loading of one rail and the unloading of the other. It is to be particularly emphasized that horizontal and vertical nonuniformities are mutually dependent, and can become stronger with time.

A vertical track heave, for example, produces a small lurching motion of the wheel-set; this causes, after the heave is passed, a pressure of the wheel-set against one rail, and indeed (for a given velocity and a given mass distribution of the vehicle) a pressure that always acts on the same location of the rail. Gradually, therefore, a horizontal buckle of the rail will be formed that corresponds to the vertical heave. If we begin, on the other hand, with a horizontal nonuniformity, then this nonuniformity causes an increased load on one rail; an increased load on the curved rail itself if the curvature is outwardly convex, and an increased loading on the other rail if the curvature is inwardly convex. As a sufficient number of wheel-sets and vehicles pass this position, the more heavily loaded rail will gradually sag or be ground down. *A vertical track heave therefore generates a horizontal buckle, and a horizontal buckle a vertical heave; the gyroscopic effect therefore works systematically for the deterioration of the track.* The combined effects of the horizontal and vertical irregularities and the coupling of the moments K and K' by means of the lurching motion can become a direct risk to operational safety; it would be well worthwhile to test this theoretical possibility by experiments.

As a more damaging circumstance, a type of *resonance effect* can come into play in cases in which the gyroscopic effects follow one another rhythmically (as, for example, do the effects of eccentric wheels or mutually dependent horizontal and vertical irregularities), and in which this rhythm coincides with the rhythm of the free oscillations of a sprung carriage.*)

The reader will probably be inclined to regard these conclusions as theoretical exaggerations if he compares them with his experience of the comfort and safety of a well-sprung railway carriage. It is to be remarked, however, that according to formulas (4) and (6), which hold not only for uniform curves and superelevations but also correspondingly for track heaves, *all gyroscopic effects increase with the square of the velocity of travel* v. If these effects may be hardly perceptible under normal circumstances, one reaches by a corresponding increase of the traveling velocity a boundary where they are perceptible, and, for a further increase, a boundary where they become dangerous. The boundaries are actually not very high, especially since a doubling of the velocity causes a quadrupling of the gyroscopic effects. In fact, it is confirmed by the participants of the high-speed trial runs in Berlin that for the original unprepared state of the rails, the feeling of safety was already transformed into the opposite at 150 km/hr. The permissible velocities in normal operation are established just so that nonuniformities in the state of the track and the wheels, as well as the oscillatory motion of the vehicles, have no harmful consequences at these speeds.

---

*) One can think of compensating for the gyroscopic effects of the wheel-set by arranging a flywheel on the same axle that rotates with oppositely equal impulse.

We thus understand why the *Studiengesellschaft für elektrische Schnellbahnen* first carried out a careful observation and strengthening of the rails before it finally proceeded to the highest obtained velocity of 200 to 210 km/hr. In particular, Mr. W i t t f e l d,*) whose publications and kind personal communications we have to thank for the viewpoint given under c), states that the upper bound for the speed of travel is determined essentially by the track construction and the rail guidance. Among the other requirements that Mr. W i t t f e l d sets for an electric or steam carriage designed for high-speed travel, we emphasize:

Wheel-sets with the smallest possible moment of inertia C, to diminish the gyroscopic effects K and K';

Vehicles with the greatest possible moment of inertia about the vertical axis through the center of gravity, to diminish the lurching motions possibly excited by gyroscopic effects.

While correcting the proofs of this section, we had occasion to see a manuscript copy of the article "Schwerpunktslage und Kreiselwirkungen bei elektrischen Lokomotiven," by Mr. G. B r e c h t, which will soon appear in the Zeitschrift für elektrische Kraftbetriebe und Bahnen.[286] The general deliberations there correspond completely to the preceding, but are also substantiated by extensive numerical computations for specific types of locomotives. We quote from the conclusions of the author, which indicate a practical insignificance of gyroscopic effects for certain types of construction:

"Certain gyroscopic effects that are partly characteristic of electric drives must undoubtedly be considered for high-speed electric locomotives. The gyroscopic effects are smallest for locomotives with gear motors, because of the different rotational senses of the rotating masses. They are larger for axle-mounted motors, under otherwise equal circumstances ...."

"The uniform gyroscopic effects that appear in curves and superelevations (cf. a) and b) above) change the perpendicular and lateral rail pressures by at most 5–15%. These effects play no role in comparison with other influences, especially in comparison with centrifugal forces and the inevitable impact effects."

"The effects caused by sprung rotating masses have somewhat greater significance for the more recent type of locomotive construction. It is possible that gyroscopic effects can cause a lurching vehicle to roll, and vice versa (cf. the conclusion of b) above). These phenomena are exceptional, however, and can occur to a degree that is of practical importance only for speeds

*) Cf. his report Über Schnellbahnen und elektrische Zugförderung auf Hauptbahnen. Annalen für Gewerbe and Bauwesen, edited by Glaser, Bd. 50, Heft 5, 1902, p. 86.[285]

greater than 200 km/hr; even then they do not impair operational safety. In addition, both oscillations are opposed by considerable frictional resistance, so that they are simultaneously brought to a rapid extinction ...."

"The use of axle-mounted motors has the consequence that the impulse of the individual wheel-sets is relatively large. If the axles of such a locomotive experience short, rapidly successive turning-impacts in their vertical plane on a particularly poor section of the track (cf. c) above), these impacts could very well cause strong horizontal rotations of the axles; there is then a real possibility that the entire undercarriage of the locomotive may receive an energetic turning-impact that produces strong lurching .... It is not out of the question that for the concurrence of various unfavorable circumstances and very poor track condition, axles with built-in fixed motor armatures can actually produce serious gyroscopic effects."

§3. The torpedo guidance apparatus. The Whitehead and Howell torpedoes.

A well-studied and practically effective example of gyroscopic stabilization is provided by the *guidance apparatus of the Whitehead torpedo.* The characteristic capability of the top to resist changes in direction, its "absolute orientation in space," as we said at the conclusion of Chap. III, is developed here to full fruition.

In order to explain the effectiveness of the guidance apparatus, it is necessary to step back somewhat, and to sketch the general arrangement of a torpedo.

The W h i t e h e a d torpedo has the approximate form of a cigar; it is 5 m long and has, at the thickest location, a diameter of 45 cm (cf. Fig. 120).[287] The head of the torpedo contains the charge (gun

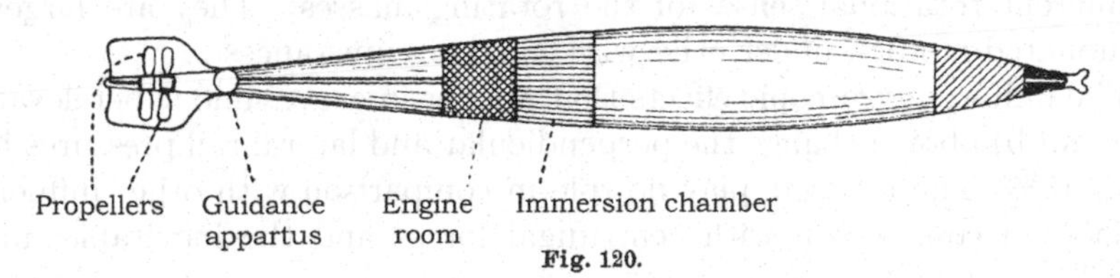

Fig. 120.

cotton) and the fuse. Behind the head is the air flask (300 liter volume), which occupies the entire middle section of the torpedo. It is filled with pressurized air at 100 atm, which is the source of all work that the torpedo expends in its course; it activates the engine that maintains the velocity of the torpedo and operates the rudders. Aft of the air flask is first the immersion chamber, which houses the *depth-control apparatus*, of which we will soon speak, and then the engine room. The output of the three-cylinder engine is about 60 horsepower; the air pressure is reduced to 30 atm before it enters the engine. The shafts of the engine run to the propellers in the tailpiece through the after-body, at whose end is interposed the *guidance apparatus*, which is of primary interest to us. The two propellers, arranged one behind the other, turn in opposite directions to avoid a net turning-moment and make approximately 1000 revolutions per minute. In addition to the propellers, the tailpiece contains the rudders for the depth-control apparatus in its horizontal fins and the rudders for the guidance apparatus in its vertical fins.

The maximum firing range of the torpedo is 600 m,*) and its speed of travel is about 15 m/sec. It is fired partly above and partly below the water. The center of gravity of the total mass of the torpedo lies 1 cm lower than the center of gravity of the mass of the displaced water, so that a certain, if small, stability against listing results.

We now describe, on the one hand, the depth-control apparatus, and, on the other hand, the guidance apparatus.

The purpose of the *depth-control apparatus* is to maintain the torpedo at a given depth (3 to 4 m beneath the surface of the water) that is set before firing. This purpose is served by a piston or disc with a membrane that is in contact with the water on one side and is placed against a spiral spring on the other side. The piston is in equilibrium if the water pressure and the spring force are equal; the spring is regulated so that this occurs at the desired depth of 3 to 4 m. If the torpedo overshoots this depth, the water pressure increases according to the well-known hydrostatic law; it is now stronger than the spring force, and pushes the piston back. If the torpedo undershoots this depth, then the spring force is stronger, and the piston advances into the water. The piston is connected to the two named rudders in the tailfins through a linkage and a valve that regulates the entrance of pressurized air to the steering mechanism. The rudders are deployed upward for an inward motion of the piston, and downward for an outward motion. The pressure of the water flowing against the rudders will elevate the front of the torpedo in the first case and depress it in the second. In this manner, the deviation from the prescribed depth is corrected in both

*) This information refers to the year 1900 (cf. the note at the end of this section). A considerably greater firing range (up to 5000 m in the English navy) has more recently been sought.

cases. The introduction of the depth-control apparatus has the consequence that the deviation is always overcorrected, so that a downward deviation is followed by an upward deviation; the overcorrection is decreased by a specially designed pendulum device that restores the rudder. On the whole, therefore, the torpedo will oscillate around the prescribed depth on a shallow wavy line.[288]

The *guidance apparatus**) was designed by the Austrian engineer Obry;[290] its purpose is to maintain a straight course of the torpedo, as determined by the sighting device at the firing. Its office is more to *note* a lateral deviation from the straight course than to *annul* or *reverse* it. The latter function falls to the pressurized air that flows from the air flask in fine tubes**) through a ventilation system controlled by the guidance apparatus, and deploys two rigidly coupled rudders in the tailpiece to the right for leftward deviations and to the left for rightward deviations. The guidance apparatus has, in a certain sense, only to give the innervation, while the actual expenditure of work—the muscular exertion, as we can say in analogy with the animal body—is performed by the pressurized air of the air flask. We name this operating principle the *principle of indirect guidance*, since the apparatus acts only to activate a large reservoir of force, and the required change of direction of the torpedo is achieved by means of the latter. We will meet the opposite *principle of direct guidance* in the Howell torpedo, for which the guidance apparatus assumes, at the same time, the functions of both the nerve and the muscle. Indirect guidance has the great advantage that it requires only small force effects that can be achieved by an apparatus of moderate mass, while direct guidance requires an apparatus whose mass must be somewhat proportional to the total mass of the torpedo to be moved. The Obry apparatus weighs less than 4 kg; its rotor rings, bearings, and ventilation system are worked with precision, as is possible only for a device of small mass.

We have already alluded many times to the gyroscopic property of the top—that is, its capacity to note a change of direction—as, for example, in the description of the Foucault experiments. The Obry apparatus is naturally based on this same property.

---

*) A more detailed description of the guidance apparatus is given by Lieutenant W. J. Sears in Engineering, Vol. 66, 1898, p. 89.[289]

**) The pressure of the air is thus decreased to about 15 atm.

The form of the top in the Obry apparatus is shown in Fig. 121. It consists of a rotor in a Cardanic suspension, completely analogous to the arrangement used by Foucault. The rotor weighs 800 gr and has a diameter of 76 mm. In its normal position, the rotation axis coincides with the long axis of the torpedo. The rotor turns in an inner ring that may rotate about a horizontal axis; this inner ring is mounted in an outer ring that may rotate about the vertical and has bearings in the body of the torpedo. The rotor is fixed to a strong conical axle that is supplied on one side with sprocket depressions. The depressions are engaged by the toothed casing of a drum spring that is tightened when the torpedo is installed in the launch tube and is released at the firing. Before the spring is tightened, an automatic regulating device places the top in its normal position, with the inner ring horizontal and the outer ring perpendicular to the long axis of the torpedo. The rotational velocity imparted to the rotor is, according to information from the manufacturer, 9000 to 10000 revolutions per minute. The direction of the generated impulse vector is, at the same time, the firing direction.*)

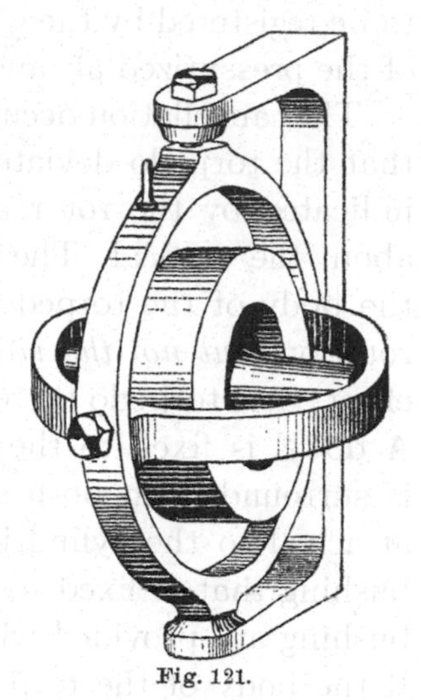

Fig. 121.

The rotor must be centered with great care, so that its center of gravity falls exactly at the intersection of the rotation axes of the inner and outer rings. Only in this case does the impulse vector retain its direction during the course of the motion. The figure axis of the rotor would otherwise change gradually according to the laws of motion of the heavy top, and describe a more or less regular precession about the vertical. The apparatus would then produce a curvature of the torpedo path, in the same manner that it directs a straight path for a centered rotor.

Significant lateral deviations from the firing direction occur particularly at the beginning of the trajectory, due to the release from the launch tube, the impingement of the water, the crossing of wave crests, and the especially strong initial action of the depth-control rudder, which not only regulates the depth of a pitching torpedo,

*) How the guidance apparatus has been adapted to the modern demands of the firing distance (cf. the note on p. 783) has not become known.

but also simultaneously steers it somewhat to the side. All these irregularities, which would strongly influence the certainty of a strike, are to be registered by the guidance apparatus, and canceled with the help of the pressurized air and the steering rudder.

The cancellation occurs in the following simple manner. We imagine that the torpedo deviates from its straight course, whose direction is indicated by the rotor axis in Fig. 121, so that the torpedo rotates about the vertical. The bracket shown in the figure, which is fixed to the body of the torpedo and bears the outer ring, takes part in this rotation, *but not the ring itself*, which is stabilized by the gyroscopic effect. The torpedo therefore rotates laterally with respect to this ring. A dowel is fixed to the ring, as indicated in the figure. This dowel is surrounded on both sides, with some clearance, by a fork that is attached to the cylindrical ventilation valve. The valve rotates in a bushing that is fixed to the body of the torpedo. The valve and the bushing are provided with fine tubes (canals for the pressurized air). If the body of the torpedo turns with respect to the outer ring, the fork strikes against the dowel; since the dowel does not yield, the fork and the attached valve will turn with respect to the valve bushing. The tubes, which originally do not pass one another and permit no entry of the pressurized air to the steering mechanism of the guidance apparatus that lies in the rear part of the torpedo, are now aligned; the pressurized air flows through the tubes and presses the piston of the steering mechanism to one side or the other, according to the sense that the valve is turned. The piston deploys the lateral rudder by means of a rod and linkage, and the lateral rudder directs the torpedo back toward its straight course, as described above.[291]

Just as for the depth-control apparatus, the lateral deviation is consistently somewhat overcorrected, since the reversal of the ventilation valve and the restoring of the rudder can result only if, for example, the rightward deviation has gone over into a leftward deviation. The horizontal projection of the trajectory, just like the vertical projection, will thus be a wavy line that oscillates about the straight direction. Since strong lateral deviations occur only at the beginning of the trajectory and the lateral oscillations produced by the guidance apparatus itself are small with respect to those initial deviations, the amplitude of the wavy line gradually decreases, and the trajectory adapts to the straight direction of the line of sight.

The practical usefulness and reliable operation of the O b r y guidance apparatus is demonstrated most clearly by the fact that it has been introduced in the torpedo service of almost all nations. Its use, admittedly, requires experience in handling and care in adjustment.

As we now consider the stabilizing process in numerical detail, we must first refer to the general developments in §1 sub IV. Here, as there, it is a matter of a rapidly spinning horizontal rotor that apparently does not yield to a moment that acts about the vertical, and in reality yields only very little. We will change the general stabilization schema of §1 only in one point in order to better connect to reality. We would, namely, no longer understand by A in equations (IVb), (IVc), and (IVe) the equatorial moment of inertia of the rotor alone, but rather the sum of this moment of inertia and the moment of inertia of the inner ring about its rotation axis (cf. Fig. 121), since the latter takes part in the elevation of the figure axis in the C a r d a n i c suspension; further, we would replace A in the left-hand sides of equations (IVa) and (V) by J, and thus combine the moments of inertia of the rotor and the inner and outer rings about the vertical axis of the latter, since the inner and outer rings also take part in the change of position of the figure axis that is determined by the angle ψ. Equation (IV) thus becomes

$$\frac{d^2\psi}{dt^2} = \frac{\Psi}{J} - \frac{N^2}{AJ}\psi,$$

and its integration under the corresponding initial conditions yields, in place of (IVf) and (IVg),

$$\psi = \frac{A\Psi}{N^2}\left(1 - \cos\frac{N}{\sqrt{AJ}}t\right),$$

$$\vartheta = \frac{\pi}{2} - \frac{\Psi}{N}t + \frac{\Psi\sqrt{AJ}}{N^2}\sin\frac{N}{\sqrt{AJ}}t;$$

there follow for the mean and maximum rotation angle, as previously,

$$\psi_m = \frac{A\Psi}{N^2} \quad \text{and} \quad 2\psi_m.$$

By this angle, therefore, the figure axis of the rotor initially gives way in the mean to the turning-moment Ψ, and with it the plane of the outer ring, while at the same time the plane of the inner ring slowly elevates, so that the capability of resistance of the outer ring is progressively diminished, and the angle of rotation is therefore gradually increased. The moment Ψ, regarded here as constant, has a twofold origin for the torpedo: a small frictional moment is applied from the bracket of Fig. 121 to the bearing of the outer ring for a lateral deflection of the torpedo if this ring actually remains fixed in space, and the ventilation valve, if it is moved by the outer ring, applies a small resistance that acts on the dowel of Fig. 121 and likewise produces a moment about the rotation axis of the outer ring.

For the following numerical calculation, we must fall back upon the previous information. The number of revolutions per minute is 10 000; the angular velocity per second ω is therefore

$$\omega = \frac{10\,000}{60} \cdot 2\pi = 1{,}05 \cdot 10^3 \ (\mathrm{sec}^{-1}).$$

We may assume that this velocity is constant, even though it is naturally gradually weakened, in reality, by friction in the bearing of the rotor. Experiments show, namely, that the rotor runs out for the relevant initial velocity in about 12 minutes, while the torpedo shot (see below) lasts only about 40 seconds.

The mass of the rotor is 800 gr, and its radius is 38 mm. Were the mass distributed entirely on the outer circumference, the moment of inertia about the centerline would be, in the absolute cgs system,

$$800 \cdot 3{,}8^2 = 11{,}6 \cdot 10^3 \ (\mathrm{gr\ cm}^2);$$

were the mass, on the other hand, uniformly distributed on the circular disk, we would have as the moment of inertia only

$$\frac{1}{2}\, 800 \cdot 3{,}8^2 = 5{,}8 \cdot 10^3 \ (\mathrm{gr\ cm}^2).$$

The actual mass distribution of the rotor is such that its moment of inertia C must be smaller than the first named value and larger than the second. We take as a mean value

$$C = 8 \cdot 10^3 \ (\mathrm{gr\ cm}^2).$$

For the impulse N there follows

$$N = C\omega = 8{,}4 \cdot 10^6 \ (\mathrm{gr\ cm}^2\ \mathrm{sec}^{-1}).$$

The moment of inertia of the rotor about an axis perpendicular to the centerline would be $C/2$ if the mass were distributed only in a plane perpendicular to the centerline; in reality, it turns out somewhat larger. For the moment of inertia A in our formula, moreover, the moment of inertia of the inner ring about its rotation axis is to be included. It is therefore all the more certain that $A > C/2$; we estimate

$$A = 5 \cdot 10^3 \quad (\text{gr cm}^2).$$

The moment of inertia J about the rotation axis of the outer ring comprises the moment of inertia of the rotor, that of the inner ring about the vertical, which in general is greater than the moment of inertia of the same about its rotation axis, and, in addition, that of the outer ring. Thus $J > A$; we take

$$J = 6 \cdot 10^3 \quad (\text{gr cm}^2).$$

Finally, the external turning-moment Ψ is still to be estimated. Because of the precise manner of construction of the device, the frictional moment in the bearing of the outer ring for a rotation of the body of the torpedo and the resistance of the ventilation valve on the dowel of the outer ring are small. We will not err in order of magnitude if we set both together at most equal to the moment of 1 gr weight at a lever arm of 1 cm, and therefore write

$$\Psi = g = \text{approximately } 10^3 \quad (\text{gr cm}^2 \text{ sec}^{-2}).$$

We thus calculate, on the basis of the just repeated formulas, the following quantities.

1. the mean deflection of the outer ring, at the same time the amplitude of its oscillation:

$$\psi_m = \frac{A\Psi}{N^2} = \frac{5 \cdot 10^3 \cdot 10^3}{(8{,}4 \cdot 10^6)^2} = 7 \cdot 10^{-8} = 0{,}015'';$$

2. the oscillation period of the outer ring, at the same time that of the periodic part of the motion of the inner ring:

$$\tau = 2\pi \frac{\sqrt{AJ}}{N} = 2\pi \frac{\sqrt{30} \cdot 10^3}{8{,}4 \cdot 10^6} = 4{,}1 \cdot 10^{-3} \text{ seconds};$$

3. the reciprocal mean angular velocity of the inner ring; that is, the time in which the inclination ϑ of the inner ring changes by a unit angle in radian measure:

$$\frac{1}{\vartheta'_m} = \frac{N}{\Psi} = \frac{8{,}4 \cdot 10^6}{10^3} = 8{,}4 \cdot 10^3 \text{ seconds},$$

and the time in which, for example, the inclination changes by 1°:

$$\frac{8{,}4 \cdot 10^3}{57{,}3} = 147 \text{ seconds} = 2\,1/2 \text{ minutes}.$$

These numbers show us the following.

1. The mean deflection of the outer ring is, in fact, entirely imperceptible; the outer ring is completely stabilized by the gyroscopic effect.

2. The oscillations of the outer ring are imperceptible not only because of the smallness of their amplitude, but also because of their extraordinarily short period. The same holds for the oscillations of the inner ring.

3. The uniform mean rotation of the inner ring, in contrast, would become perceptible in a not too short observation time; in the available time of the torpedo shot, however, this motion is hardly significant. We said that the torpedo runs with a velocity of 15 m/s and that the maximum firing range is 600 m. The duration of the shot will therefore be at most 40 sec; for this time, our calculation above yields a deflection of the inner ring by

$$\frac{40^\circ}{147} = 16';$$

the inner ring thus retains almost precisely its original normal position. For this small deviation, the capacity of resistance of the outer ring remains almost unweakened, so that the application of our approximate formulas, which according to §1 sub IV would fail for larger deviations, is justified here. It is to be added that our above assumption of a temporally constant moment Ψ does not correspond to the actual conditions, and represents the most unfavorable case imaginable. In reality, the moment Ψ changes because of the counteraction of the rudder, and changes its sense many times during a shot because of the resulting periodic character of the lateral deviations. With the reversal of the moment Ψ, however, the rotational sense of the inner ring also reverses; the angular change ϑ of the inner ring will in this manner again be made partly retrograde. Thus the inner ring will also appear at the end of the shot in almost exactly its normal position at the firing.

It is noted that depth deviations of the torpedo have a certain, if small, influence on the effectiveness of the guidance apparatus. If the body of the torpedo inclines upward or downward, the rotation axis of the outer ring will take part and turn with respect to the vertical by the same angle as the centerline of the body of the torpedo. In the first approximation, however, the inner ring does not take part in this rotation, since it moves freely in its rotation axis and is held fixed in its original horizontal position by the gyroscopic effect of the rotor; only to the extent that a frictional moment is transmitted to the inner ring by the inclination of the outer ring will a small mean deflection of the inner ring occur, and in conjunction with it a slow rotation of the outer ring. The roles of the inner and outer rings are now reversed with respect to the previous consideration, in that the primary frictional moment Ψ now acts on the axis of the inner ring. A perceptible rotation of the outer ring is again not caused, since the depth deviations also change periodically. We mention this characteristic coupling of the rotations of the inner and outer rings, or the coupling of the depth and lateral oscillations, primarily with respect to the next considered *Howell torpedo.*

The theoretical deliberation therefore explains the favorable experiments that have been made with the Obry apparatus.

We must finally present, as a counterpart to the indirect guidance apparatus of the Whitehead torpedo, the *direct guidance apparatus of the Howell torpedo.*[292] Here too we must begin with a general description of the torpedo.

The external form and dimensions are approximately those of the Whitehead torpedo: length 4,27 m, maximum diameter 45 cm, total weight 520 kg. Approximately in the middle of the torpedo lies a strong transversely mounted flywheel of weight 133,8 kg, which therefore makes up an essential fraction of the total weight. The plane of the flywheel coincides with the vertical midlength plane of the torpedo. Before the firing, the flywheel is given an angular velocity of 9500 revolutions per minute by means of a steam turbine, so that a formidable reservoir of work is stored in the significant mass of the flywheel. This is the single source of energy for the torpedo, by which the velocity of the forward motion is maintained and the direction of the velocity is corrected. *The rotating flywheel serves in the Howell torpedo as both the propulsive engine and the guidance apparatus.*

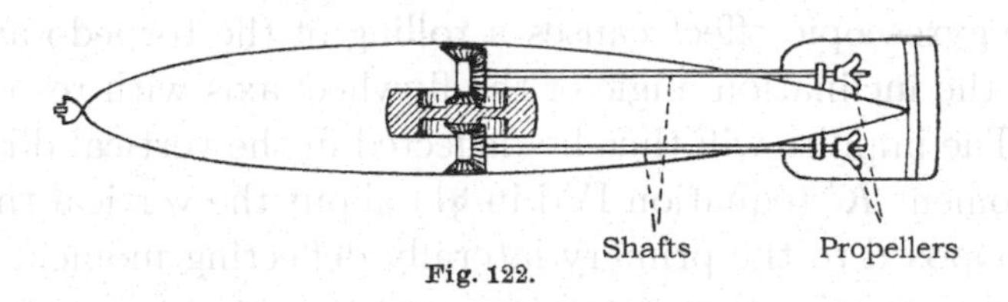

Fig. 122.

The rotation of the flywheel is transmitted directly to the shafts of the propellers by two pairs of conical gears (cf. Fig. 122) in a 2:1 transmission ratio. The pitch of the propellers is variable, and is increased with increasing distance of travel of the torpedo; the velocity of the torpedo thus remains approximately constant, in spite of the considerable decrease of the flywheel rotation speed due to the expenditure of external work.

In order to judge the intrinsic directional force of the flywheel, we first ask for the kind and number of its degrees of freedom. If the body of the torpedo is imagined to be fixed, the flywheel has only one degree of freedom, its own simple rotation. The axis of this rotation lies here transversely, while for the Obry apparatus the corresponding axis of the rotor in the normal position is disposed fore and aft. We have a second degree of freedom in the rotation of the torpedo about its long axis—the rolling of the torpedo—through which the axis of the flywheel is inclined with respect to the horizontal. This rolling replaces, in a certain manner, the degree of freedom of the inner ring in the Obry apparatus. Since the Howell torpedo—just like the Whitehead torpedo—possesses a certain stability (here too the center of gravity of the torpedo lies somewhat beneath that of the displaced water mass), the rolling will generally be periodic, and oscillations to one side and the other will result. As the third degree of freedom, corresponding to the outer ring of the Obry apparatus, we must regard the rotation of the torpedo about the vertical, which occurs in lateral deviations. Rotations about the transverse horizontal axis, which are associated with depth deviations, do not, in contrast, influence the position of the flywheel, and thus fall out of consideration for us.

We must imagine the formation of the gyroscopic effect for the Howell torpedo in the following sequence of events. Let a lateral deviation due to an external influence be the primary cause. This would turn the impulse of the flywheel horizontally; the direction of the impulse change is thus the long axis of the torpedo. The resulting countermoment of the gyroscopic effect K, given by equation IVb in §1, acts about this axis. The gyroscopic effect causes a rolling of the torpedo and thus a change of the inclination angle of the flywheel axis with respect to the vertical. The impulse will thus be deflected in the vertical direction. A countermoment K' (equation IVd in §1) about the vertical thus arises, which is opposed to the primary laterally deflecting moment.

We thus recognize that a correction of the lateral deviation by the gyroscopic effect occurs again here, and that it is mediated by the rolling of the torpedo. One should not overlook the fact that this rolling is less unhindered than the corresponding rotation of the inner ring in the Obry apparatus, and that water resistance acts against the rolling in a significant measure. In that it does so and the rolling velocity is thus decreased, the final moment K', whose magnitude was found to be proportional to the rolling velocity ϑ', is also decreased. Because of water resistance and the coinciding effect of gravity, through which the sense of the rolling is periodically reversed, the circumstances here are not as favorable as for the carefully adjusted Obry apparatus, which is completely surrounded by the torpedo hull.

We can also, however, reverse the sequence of the above deliberation, and regard a rolling of the torpedo as the primary cause. In striking the water, namely, this would be generated in a significant measure by external influences. Exactly in the manner described above, a moment about the vertical would result from the rolling by mediation of the gyroscopic effect of the flywheel, and thus also a *lateral deviation.* We could conclude through a continuation of the argument that a moment acts against the rolling, but the water resistance against lateral deviations of the extended form of the torpedo is so large that this countermoment is hardly worth mentioning. In addition, it is not the hindering of the rolling that interests us, but rather the hindering of the lateral deviations. With respect to the latter, however, our deliberation shows that *the flywheel of the Howell torpedo not only appears to be a device that hinders lateral deviations or reverses them, but rather, at the same time, also produces them on the basis of otherwise generated rolling.* The characteristic coupling of these two degrees of freedom of the rotor, to which we called attention above, is clearly expressed here (rolling and lateral deviations for the Howell torpedo, lateral and depth deviations for the Whitehead torpedo).

The manufacturer of the Howell torpedo appears to recognize that our apprehension about the damaging influence of rolling motion is justified, for he introduces vertical rudders that are intended to ensure the upright position of the torpedo. As an objection to this arrangement, however, one cannot but present the following alternative: *either the rudders do not hinder the rolling, and thus the rolling generates lateral deviations, or the rudders hinder the rolling, and thus they also hinder the directional force; the rolling is a necessary intermediary for the production of the countermoment against the lateral deviations.* If the named rudders are actually effective, then the torpedo is a top with only *two* degrees of freedom; such a top possesses, according to the general principle stated in §1 on p. 768, no capability of resistance against changes of direction.

We concur, in this critique of the particularly interesting and bold Howell construction, with the statements of Mr. Diegel,*) in whose work, on the other hand, our preceding considerations on the Obry apparatus are taken up. We refer to the presentation of this expert for further details.[293]

The Howell torpedo does not appear to have proven of value in practice; at least the government of the United States, which had purchased the Howell torpedo, later went over to the Whitehead torpedo with its indirect guidance apparatus.[294]

§4. The Schlick gyroscopic ship stabilizer. Generalities and theory.

At the beginning of the seventh decade of the preceding century, the English engineer Bessemer, famous for his work in the steel industry, built a ship with a saloon cabin for sailing the English channel, with the intention that rolling motions of the cabin (oscillations about the long axis of the ship) should be prevented by the gyroscopic effect. The cabin was suspended on two strong bearings, so that it could oscillate about a horizontal axis running fore and aft, and originally carried a rigidly mounted flywheel.[295] Here, however, the essential principle of any gyroscopic stabilization (cf. p. 767) was overlooked: *the Bessemer rotor lacked one degree of freedom, and thus any possibility of influencing the rolling motion of the cabin.***) The intended gyroscopic stabilization was then replaced by hydraulic machinery,***) which, however, was no longer automatic, but rather was actuated by hand. It appears that this too did not work satisfactorily; in any case, the problem of ship stabilization was first taken up successfully thirty years later by Otto Schlick in Hamburg, whom shipbuilding already

*) Selbstthätige Steuerung der Torpedos durch den Geradlaufapparat. Marine-Rundschau 5. Heft 1899. The manuscript of this section originated at approximately the same time; possible newer experiments in the torpedo service are therefore not considered.

**) Cf. Institution of naval architects, April 1889, Remarks by Macfarlane Gray in the discussion of a presentation by Beauchamp Tower.[296]

***) Institution of naval architects, March 1875, article by E. J. Reed, on the Bessemer Steam Ship; cf. also Engineer, article by Bessemer of the same date.

owed the development of mass balancing for ship engines, and thus the possibility of modern ships with increased size and load capacity.[297]

The *Schlick ship stabilizer* has, in the first place, the required degrees of freedom. The rotor of the stabilizer turns not about an axis with bearings fixed in the ship, but rather is borne in a frame that can oscillate about a transverse axis, so that the figure axis of the rotor oscillates in the vertical fore–aft plane of the ship. The design for the steamship "Silvana" of the Hamburg-America line is represented schematically in our Fig. 123. The axis of the rotor stands vertically in the figure. In operation, it oscillates back and forth like a pendulum about this position; it is drawn back to the vertical position by a counterweight G. The rotor frame R with its horizontal rotation axis ZZ is seen in the figure; it carries the rotor, which is designated by S. The rotor has a diameter of 1,6 m and a weight (without its axle) of 5100 kg. It is given a rotation number of 1800 turns per minute, corresponding to a circumferential velocity of 150,8 m. The rotor is driven like a steam turbine, the steam entering through one journal of the frame and escaping through the other. The requirement of the large circumferential speed naturally demands the most careful construction and the best material (a circumferential speed of only 30 to 40 m is ordinarily allowed for cast iron flywheels). The installation was designed by Dr. F ö t t i n g e r and executed by the Stettin firm "Vulkan." The stabilizer is placed near the middle of the ship.[298]

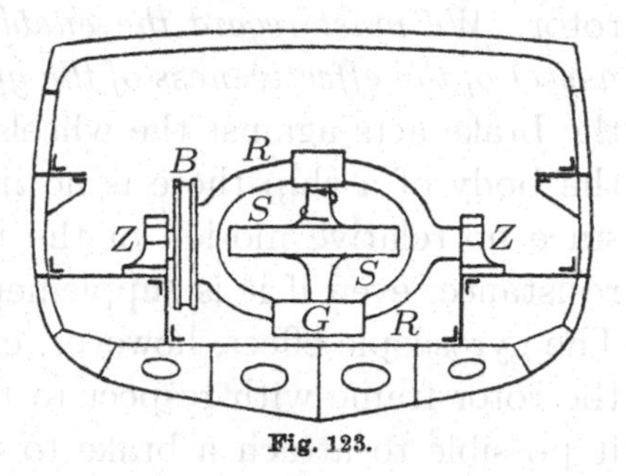

Fig. 123.

In addition to the "Silvana," the English steamship "Lochiel," a steamship of the state of Hamburg, and the discarded torpedo boat "Seebär" of the German Navy were also fitted with stabilizers, the latter for the execution of the initial orienting experiments. In addition, a stabilized ship was built in Danzig.

We first describe the general purpose of the ship stabilizer. For the prevention of sea sickness and the facilitation of firing guns on warships, it is necessary to diminish the amplitude of the rolling motion generated by ocean swells, and to increase the damping and possibly the period of the free rolling motion that follows an initially generated lateral motion; to combat the oscillations of the ship about its transverse axis, its "pitching," lies outside the office of the gyroscopic ship stabilizer.

The energy imparted to the ship in a rolling motion can naturally not be annihilated immediately, but can only be converted into an oscillation of the rotor frame with respect to the ship. In this form, however, the oscillation energy may be eliminated (that is, changed into heat) by an appropriate braking mechanism. A band brake (B in Fig. 123) and a hydraulic brake cylinder are used, both acting against the frame of the rotor. *We must regard the enablement of this braking as an essential aspect of the effectiveness of the gyroscopic ship stabilizer.* In a carriage, the brake acts against the wheels that turn relative to the chassis. In the body of a ship there is no initial possibility of applying a brake, since no relative motion of the rigid parts is present, and the water resistance, even if it is supplemented by a bilge keel, is not sufficient. The gyroscopic effect, however, creates an energetic relative motion of the rotor frame with respect to the body of the ship, and thus makes it possible to attach a brake to the latter. The brake may naturally not be so strong that it completely prevents the motion of the frame; otherwise we would again fall back on an ineffective rotor borne rigidly in the body of the ship. It is shown by experiments and the concurring theoretical deliberations that there is an appropriate mean measure of the braking that not only damps the free oscillations of the ship sufficiently strongly without disabling the gyroscopic effect, but also suppresses the forced oscillations.

We will report in §6 on experiments with the gyroscopic ship stabilizer. We first turn to its theory. It may be remarked in advance that we can represent the configuration of the stabilizer schematically by the torpedo guidance apparatus in Fig. 121. The outer ring corresponds to the body of the ship, and the inner ring to the frame of the rotor. The difference is, in the first place, only qualitative: the outer ring in the case of the ship stabilizer is to be imagined as extraordinarily large in proportion to the other parts. That its rotation axis is now horizontal rather than vertical is naturally irrelevant. For the free oscillations to be considered first, the external moment Ψ is zero, and for the oscillations that are forced by the ocean swells, the external moment Ψ changes periodically. There is, however, a further difference that is convenient to note, in order to take a different starting point in the development of the formulas: we must now introduce the named frictional effects, which we did not have to consider previously, as essential to the theory; further, we must include an uprighting gravitational moment in the calculations for the ship and the stabilizer, while we previously eliminated gravity by imagining the inner ring, outer ring, and rotor to be precisely centered. We begin by considering *two coupled oscillatory degrees of freedom*: one degree of freedom is formed by the rolling motion of the ship about its long axis, and the other by the oscillations of the rotor frame about the transverse axis of the ship; the coupling of the two degrees of freedom is due to the gyroscopic effect acting between them. We can ignore the degree of freedom of the rotor about its own figure axis, since its impulse is maintained at a constant value by the steam drive.

Strictly speaking, the stabilizer has additional degrees of freedom due to the pitching of the ship and its possible rotation about the vertical. The latter, however, will be small because of the large water resistance opposing it; the pitching motion of the ship, on the other hand, will be carried over to the oscillating energy of the rotor frame in only a small measure because of the large moment of inertia about the transverse axis of the ship. We therefore consider, for the present, the ship as fixed except for its rolling motion; we will, however, return to the pitching motion once again in §6 and estimate its influence numerically (see p. 841).

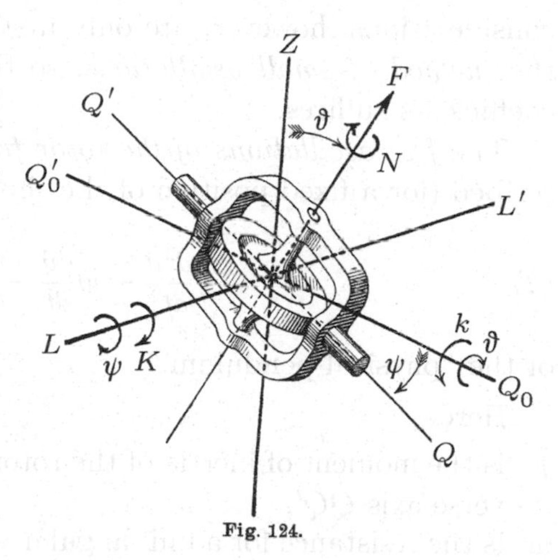

Fig. 124.

Let ψ be the angle of the rolling motion of the ship about its long axis LL' (Fig. 124), and ϑ the rotation angle*) of the frame about its transverse axis QQ', measured from the mean upright positions of the ship and the rotor frame, respectively.

The mean horizontal position of the rotation axis QQ' is denoted in the figure by Q_0Q_0', which corresponds to the angle $\psi = 0$.

We first consider the two degrees of freedom in themselves, without

*The designations ψ and ϑ should recall the Euler angles. In fact, the long axis LL' plays here the same role as the vertical in §1, about which ψ is measured. It is, however, advantageous to designate by ϑ not the angle between LL' and the figure axis, as previously, but rather the complement of this angle.

regard to their coupling. The *free oscillations of the ship*, for a fixed rotor frame, are described in a well-known manner by the differential equation

$$(1) \qquad J\frac{d^2\psi}{dt^2} + W\frac{d\psi}{dt} + QH\psi = 0.$$

Here

J is the moment of inertia of the ship about the long axis LL',
W is the resistance of the water for a unit angular velocity,
Q is the weight of the ship or the displaced water (*déplacement*),
H is the so-called metacentric height, so that QH is the uprighting moment for a unit angle of rotation.*)

If we regard the coefficients in this equation as constants, then it is known to be correct only in the first approximation (for sufficiently small ψ), since H is, in reality, a complicated function of ψ, and W is a somewhat unknown function**) of $d\psi/dt$. All of the following considerations, however, are only first approximations in the sense of the *method of small oscillations*, so that the assumption of constant coefficients suffices.

The *free oscillations of the rotor frame*, on the other hand, are described (for a fixed position of the ship) by the differential equation

$$(2) \qquad j\frac{d^2\vartheta}{dt^2} + w\frac{d\vartheta}{dt} + qh\vartheta = 0$$

of the "physical pendulum."

Here

j is the moment of inertia of the rotor and the frame about the transverse axis QQ',
w is the resistance for a unit angular velocity of the frame, which comprises the (not inconsiderable) bearing resistance and the braking resistance applied to the circumference of the frame,
q is the weight of the rotor frame and the rotor,
h is the distance of the total center of gravity of the rotor and the frame from the rotation axis QQ'.

*) For more details, see Encykl. d. Math. Wiss. IV, Art. 6 by Stäckel, Nr. 39.[299]

**) More details are found in Encykl. d. Math. Wiss. IV, Art. 22 by Kriloff and Müller, Nr. 8.[300]

The periods T and τ of the free oscillations of the two degrees of freedom, which we wish to calculate under the neglect of frictional resistance, follow from (1) and (2) as

$$T = 2\pi\sqrt{\frac{J}{QH}}, \quad \tau = 2\pi\sqrt{\frac{j}{qh}},$$

so that the corresponding "frequencies" are

$$\frac{2\pi}{T} = \sqrt{\frac{QH}{J}}, \quad \frac{2\pi}{\tau} = \sqrt{\frac{qh}{j}}. \text{*)} \tag{3}$$

Our pendulum equation (2) holds independently of whether the rotor spins. In fact, the gyroscopic effects of the rotor influence the motion, according to the often used principle of §1, only when the rotor has an additional degree of freedom. As long as the rotation axis QQ' is assumed to remain fixed, the gyroscopic effect is manifested only as a pressure on the bearings, and in no way influences the motion of the pendulum. It comes into operation, however, if we now consider the two degrees of freedom simultaneously.

If the rotor axis OF turns in the sense of positive ϑ (cf. Fig. 124), there arises a gyroscopic effect about the axis perpendicular to OF and OQ, in such a sense that it tends to place the figure axis parallel to the axis of this added rotation, and thus depresses the semiaxis OQ. The bearing on the left-hand side of the transverse axis in the figure (at Q') thus experiences an upward force, and the right-hand bearing a downward force. It is these bearing forces, that must be taken up by the rigidity of the ship, which now act as a motion-producing moment. With consideration of the just established sense, which corresponds to the positive sense of ψ, the gyroscopic effect on the moving ship is, according to equation (I) of §1,

$$+N\frac{d\vartheta}{dt}. \tag{4}$$

The eigenimpulse N of the rotor is determined in our case as the product of the moment of inertia of the rotor (without its frame) about its axis and the angular velocity produced by the turbine drive.

If, on the other hand, the ship rotates in the direction of increasing ψ, then there arises a gyroscopic effect about the axis OQ perpendicular to OF and OL. It again seeks to place the figure axis in equi-

*) In this often used designation, "frequency" denotes the number of oscillations in the unit time of 2π seconds.

orientational parallelism with the axis OL, and thus acts in the direction of decreasing ϑ. According to equation (I) of §1, the magnitude and sense of this gyroscopic effect are given by

$$-N\frac{d\psi}{dt}. \tag{4'}$$

These two moments act as external causes of motion, and are, as such, to be added to the right-hand sides of equations (1) and (2), respectively.

It is to be remarked that the application of equation (I) in the calculation of the gyroscopic effect (4′) is justified only as long as the angle between OF and OL is approximately a right angle, and therefore ϑ is small. The more precise equation (II) would give, instead of (4′), the value $-N\cos\vartheta\,\frac{d\psi}{dt}$ (ϑ now signifies the complement of the previous ϑ, and thus the sine function is replaced by the cosine), which, however, differs from (4′) only by small quantities of higher order. Similarly, it is to be noted that the gyroscopic effect (4) actually acts not about OL, but rather about the common perpendicular to the axes OQ and OF. Since this axis forms the angle ϑ with OL, the factor $\cos\vartheta$ would likewise enter in (4); this factor, however, is again to be set equal to 1 in the sense of the method of small oscillations. Finally, one can remark that a rotation of the rotor frame also influences the motion of the ship by changing its moment of inertia J. But disregarding that this change is also measured only by the higher powers of the assumed small angle ϑ and is thus, according to the method of small oscillations, likewise to be neglected, the mass of the stabilizer is so vanishingly small with respect to that of the entire ship that this influence does not at all come into question.

With the added terms (4) and (4′), equations (1) and (2) are now transformed into the *equations of motion*

$$J\frac{d^2\psi}{dt^2}+W\frac{d\psi}{dt}+QH\psi=+N\frac{d\vartheta}{dt},$$

$$j\frac{d^2\vartheta}{dt^2}+w\frac{d\vartheta}{dt}+qh\vartheta=-N\frac{d\psi}{dt}$$

for the free oscillations of the system composed of the ship and the stabilizer. We now have the *coupled* (simultaneous) *system of two linear differential equations with constant coefficients*

$$\left\{\begin{aligned} &J\frac{d^2\psi}{dt^2}+W\frac{d\psi}{dt}-N\frac{d\vartheta}{dt}+QH\psi=0,\\ &j\frac{d^2\vartheta}{dt^2}+w\frac{d\vartheta}{dt}+N\frac{d\psi}{dt}+qh\vartheta=0.\end{aligned}\right. \tag{5}$$

The coupling is a *velocity coupling*, since it is mediated by terms with the first differential quotients. It is, moreover, of a *conservative* nature, since it can produce no energy dissipation, but only an energy exchange. This is expressed analytically by the opposite equality of the "coupling coefficients" $(+N, -N)$ in (5), and is proven by the formation of the energy equation in the following manner. One multiplies the two equations (5) by $\frac{d\psi}{dt}$ and $\frac{d\vartheta}{dt}$, respectively; the coupling terms then cancel in the sum, and there follows

$$\frac{1}{2}\frac{d}{dt}\left\{J\left(\frac{d\psi}{dt}\right)^2 + j\left(\frac{d\vartheta}{dt}\right)^2\right\} + \frac{1}{2}\frac{d}{dt}\left\{QH\psi^2 + qh\vartheta^2\right\}$$
$$= -W\left(\frac{d\psi}{dt}\right)^2 - w\left(\frac{d\vartheta}{dt}\right)^2.$$

The two terms on the left-hand side signify the rates of change of the kinetic and potential energy of the system. The two terms on the right-hand side can be designated, according to Lord Rayleigh, as the dissipation function; they depend only on the original frictional terms, and are not influenced by the coupling.

We have here, just as on p. 766, an example of the *gyroscopic terms**) introduced by Kelvin; the Kelvin designation is elucidated by our case, as our coupling terms are evident signs of the inner rotation of the system that produces the gyroscopic effects.

We turn to the integration of the system (5). This is accomplished in a well-known manner by assuming

$$\psi = Ae^{i\alpha t}, \quad \vartheta = ae^{i\alpha t}.$$

Equations (5) yield for A, a, and α the conditions

$$A(-J\alpha^2 + iW\alpha + QH) = +\,aiN\alpha,$$
$$a(-j\alpha^2 + iw\alpha + qh) = -\,AiN\alpha,$$

or

$$\text{(6)} \qquad \frac{A}{a} = \frac{+iN\alpha}{-J\alpha^2 + iW\alpha + QH} = \frac{-j\alpha^2 + iw\alpha + qh}{-iN\alpha}.$$

For α there follows the equation of the fourth degree

$$\text{(7)} \qquad (J\alpha^2 - iW\alpha - QH)(j\alpha^2 - iw\alpha - qh) = N^2\alpha^2.$$

Its four roots α_1, α_2, α_3, α_4 correspond, according to (6), to the four

*) Cf. the end note at the conclusion of §4.

ratios of the magnitudes $A_1 : a_1$, $A_2 : a_2$, $A_3 : a_3$, $A_4 : a_4$. The general solution of equation (5) is now written as

$$(8) \qquad \begin{cases} \psi = A_1 e^{i\alpha_1 t} + A_2 e^{i\alpha_2 t} + A_3 e^{i\alpha_3 t} + A_4 e^{i\alpha_4 t}, \\ \vartheta = a_1 e^{i\alpha_1 t} + a_2 e^{i\alpha_2 t} + a_3 e^{i\alpha_3 t} + a_4 e^{i\alpha_4 t}. \end{cases}$$

Of the 8 coefficients present here, 4 remain, according to equation (6), arbitrary; they are sufficient to adapt the general solution to an arbitrary initial state.

(For the purpose of a later application, we wish to work out an example with specially chosen initial states in more detail.

a) We imagine that the ship is initially turned about its long axis by an angle ψ_0, with the frame of the rotor at rest in its vertical equilibrium position. The initial conditions for $t = 0$ are then

$$\psi = \psi_0; \quad \frac{d\psi}{dt} = \vartheta = \frac{d\vartheta}{dt} = 0.$$

We wish to disregard the damping here, and therefore set $w = W = 0$. Then, according to (7), $\alpha_3 = -\alpha_1$, $\alpha_4 = -\alpha_2$, and one easily recognizes that the corresponding solution results from (6) and (8) as*)

$$\psi = A_1 \cos \alpha_1 t + A_2 \cos \alpha_2 t,$$
$$\vartheta = a_1 \sin \alpha_1 t + a_2 \sin \alpha_2 t,$$

where

$$(8') \qquad a_1\alpha_1 + a_2\alpha_2 = 0, \quad A_1 + A_2 = \psi_0.$$

From (6) and (8′) then follows the ratio

$$(8'') \qquad \left|\frac{A_1}{A_2}\right| = \frac{(j\alpha_1^2 - qh)\cdot(QH - J\alpha_2^2)}{N^2\alpha_1^2}$$

in which the two oscillations contribute to the motion of the ship.

b) If we had, in contrast, taken the ship initially at rest and turned the frame of the rotor by an angle ϑ_0, then we would have obtained the entirely analogous representation

$$(8''') \qquad \begin{cases} \psi = A_1 \sin \alpha_1 t + A_2 \sin \alpha_2 t, \\ \vartheta = a_1 \cos \alpha_1 t + a_2 \cos \alpha_2 t, \\ A_1\alpha_1 + A_2\alpha_2 = 0, \quad a_1 + a_2 = \vartheta_0, \\ \left|\dfrac{a_2}{a_1}\right| = \dfrac{(j\alpha_1^2 - qh)(QH - J\alpha_2^2)}{N^2\alpha_2^2}. \end{cases}$$

We now understand by α_1 the root that is transformed for vanishing coupling into the frequency of the free ship oscillation and α_2 as the root that is transformed into the free stabilizer oscillation (see eqn. (3)),

*) Cf. the analogous representation on p. 368 for the upright motion of the top that is changed by an impact.

and distinguish these as the "primary oscillation" and the "secondary oscillation." If we exclude the case of resonance between the two oscillations, and therefore assume that $QH/J \neq qh/j$, it follows in case a), because of the large moment of inertia J, that the amplitude A_1 is much larger than A_2 for all practically achievable values of the impulse N, and in case b), in contrast, that the amplitude a_2 is much larger than a_1, while in the first case the ratio a_2/a_1 and in the second case the ratio A_1/A_2 differ relatively little from unity. Thus if the ship (case a)) gives rise to the oscillations, the primary oscillation predominates in the motion of the ship, while the two oscillations are somewhat evenly distributed for the stabilizer. The corresponding interchange of behavior occurs if the rotor (case b)) gives rise to the oscillations.

On the other hand, according to formula (6),

$$\left|\frac{A_2}{a_2}\right| = \left|\frac{N\alpha_2}{-J\alpha_2^2 + QH}\right| \ll 1;$$

thus the secondary oscillation transmitted from the stabilizer to the ship will have only a small amplitude in case b), excluding approximate resonance. The corresponding formula for the primary oscillation transmitted from the ship to the stabilizer in case a) is

$$\left|\frac{a_1}{A_1}\right| = \left|\frac{N\alpha_1}{-j\alpha_1^2 + qh}\right|;$$

here the much smaller quantities j, qh take the places of the quantities J, QH, so that the ratio a_1/A_1 for the primary oscillation will be considerable for the practically present values of N.

Since we found above that "induced" oscillations in the other degree of freedom are approximately uniformly distributed in both frequencies for oscillations excited by the ship or the stabilizer, it follows that a) a rolling motion of the ship can excite the stabilizer relatively strongly, and b) large oscillations of the stabilizer can cause, in contrast, only small oscillations of the ship except in the case of resonance, for which care must be taken to provide sufficient braking of the secondary oscillation).

The detailed numerical discussion of equation (7) will occupy us in the following section. In preparation, it is necessary to rewrite the equation so that only *dimensionless quantities* are present. One can proceed, for example, in the following manner.

Let α_0 be the frequency of the frictionless uncoupled ship oscillation, determined by (3) as

$$\alpha_0^2 = \frac{QH}{J}, \quad \text{and let } \beta = \frac{\alpha}{\alpha_0}; \tag{9}$$

further, let v be the ratio of the frequency of the frictionless, uncoupled oscillation of the stabilizer*) to the frequency of the oscillation of the ship, so that

$$(9') \qquad v^2 = \frac{qh}{j} : \frac{QH}{J}, \text{ and thus } \frac{qh}{j} = v^2\alpha_0^2.$$

If one now introduces the likewise dimensionless abbreviations**)

$$(9'') \qquad K = \frac{W}{J\alpha_0}, \quad k = \frac{w}{j\alpha_0}, \quad n^2 = \frac{N^2}{Jj\alpha_0^2},$$

then there follows, from (7), the equation

$$(10) \qquad (\beta^2 - iK\beta - 1)(\beta^2 - ik\beta - v^2) = n^2\beta^2$$

for the now dimensionless unknown β.

Of the quantities introduced here, n measures the gyroscopic effect, k the braking effect, and K the original damping of the ship oscillation; K can be neglected under most circumstances with respect to k. We will also give, as a special case, the value 1 to the frequency ratio v; this is case in which the original oscillations of the ship and the frame of the rotor are in resonance. For $n = 0$, the roots β separate into two pairs that correspond to the uncoupled oscillations of the ship and the stabilizer. We will have to investigate below how these pairs change continuously for increasing n and appropriately chosen k. The oscillation mode that arises from the original oscillation of the ship, which we called the "primary oscillation," will have, under favorable circumstances, increased damping and an increased oscillation period, as was stipulated as desirable at the beginning of this section.

It is not, however, the *free oscillations* of the ship and the stabilizer that are of primary interest in practice, but rather the oscillations that are *forced* by the ocean swells. The fraction to which the amplitude of the latter can be reduced by the stabilizer provides the simplest, if not the single, measure for the effectiveness of the construction.

We first write the equation of motion for the ship with a fixed rotor frame on a moving sea. This is accomplished by the following

---

*) One can naturally speak of a ratio of oscillation frequencies only as long as both oscillations are periodic. In contrast, our general definition in (9′) also retains its meaning in the aperiodic case. If v is nevertheless often called a frequency ratio, this is to be understood for the aperiodic case in the generalized sense, and is to refer to the actual definition (9′).

**) The damping constant K naturally has nothing to do with the elsewhere denoted gyroscopic effect K.

approximate deliberation:*) while the uprighting moment on a smooth sea tends to align the midplane of the ship with the vertical and is thus (see equation (1)) to be set proportional to the angle ψ itself, it acts on a moving sea so that it rotates that plane into a direction that is inclined to the vertical and is approximately perpendicular to the undulating water surface, but always oscillates about the vertical with the same period as the water waves and with a magnitude that depends on the wave amplitude. The uprighting moment is now to be set proportional to $\psi - \psi_0$, where ψ_0 signifies the variable inclination of the named equilibrium position of the ship with respect to the vertical. We therefore insert $\psi - \psi_0$ in place of ψ in the third term of equation (1), while we retain the first and second terms unchanged. (The inertial effect of the rolling ship is indeed determined as before by the acceleration of the ship alone, and the water resistance is determined, at least for the most part, by the angular velocity of the ship with respect to the entire water mass, and not essentially with respect to its changing surface.) Equation (1) is thus transformed into

$$J\frac{d^2\psi}{dt^2} + W\frac{d\psi}{dt} + QH(\psi - \psi_0) = 0.$$

We assume a uniform harmonic train of water waves, which we imagine as rolling laterally against the ship. If ω is the frequency of the waves (equal to 2π divided by the oscillation period), then ψ_0 can be taken as proportional to $\cos\omega t$, and thus

$$J\frac{d^2\psi}{dt^2} + W\frac{d\psi}{dt} + QH\psi = C\cos\omega t \tag{11}$$

will be assumed. The waves on the surface of the water thus act on the ship like an external moment $C\cos\omega t$ that alternates with the period of the waves; its amplitude C increases with the amplitude of the water waves. The form of the water is naturally not, in general, a uniform sine or cosine curve. We can, however, always decompose an arbitrary periodic form into a series and treat of each term individually.

Just as equation (11) arises from equation (1) by the addition of the term $C\cos\omega t$, we obtain, from equations (5) for the free oscillations of our coupled system, the following *differential equations***) for the oscillations of the same system forced by the ocean swells*:

*) For more detailed developments and a literature review, see Encykl. d. Math. Wiss. IV, Article 22 by Kriloff and Müller, No. 4d.[301]

**) In association with p. 796, we point out that these equations are transformed for $W = w = H = h = 0$ into equations (IVb) and (IVe) of §1, if we assume there the special case $\Psi = C\cos\omega t$.

$$(12)\quad \begin{cases} J\dfrac{d^2\psi}{dt^2}+W\dfrac{d\psi}{dt}-N\dfrac{d\vartheta}{dt}+QH\psi=C\cos\omega t,\\ j\dfrac{d^2\vartheta}{dt^2}+w\dfrac{d\vartheta}{dt}+N\dfrac{d\psi}{dt}+qh\vartheta=0,\end{cases}$$

in which neither the motion of the rotor frame nor its gyroscopic effect is changed directly by the external moment.

The integration of these equations is also accomplished according to a well-known method in which we best calculate, as above, with complex quantities.

We write

$$\psi = Ae^{i\omega t}, \quad \vartheta = ae^{i\omega t},$$

and also replace the right-hand side of the first of equations (12) by the complex expression

$$Ce^{i\omega t}.$$

If the complex coefficients A and a are determined by the conditions

$$(13)\quad \begin{cases} A(-J\omega^2+iW\omega+QH)-iaN\omega=C,\\ a(-j\omega^2+iw\omega+qh)+iAN\omega=0,\end{cases}$$

which follow from (12), then our complex assumption satisfies not only the equations obtained from (12) by the insertion of $e^{i\omega t}$ in place of $\cos\omega t$, but also, at the same time, the real part of the same, the original equations. From (13) there follows

$$(14)\quad \begin{cases} A=C\dfrac{-j\omega^2+iw\omega+qh}{(-J\omega^2+iW\omega+QH)(-j\omega^2+iw\omega+qh)-N^2\omega^2},\\ a=C\dfrac{-iN\omega}{(-J\omega^2+iW\omega+QH)(-j\omega^2+iw\omega+qh)-N^2\omega^2}.\end{cases}$$

The absolute values of these quantities are the amplitudes of the ship and stabilizer oscillations that are forced by the ocean swells; the arc-tangents of the ratios of the imaginary and real parts give the phases.

One obtains the general solution of (12) from this particular solution if one superposes the general solution (8) of the differential equations for the free oscillations. Since the latter are damped, and even intentionally braked for the ship stabilizer, there remain after a short time, in the case of a regular train of waves, only our forced oscillations, whose amplitude therefore provides us with a simple measure of the effectiveness of the ship stabilizer. That this, as we said, is not the single measure, is likewise clear. The further condition must be added that the free oscillations, which will always be generated anew because of the irregular character of the amplitude, phase, and period of the ocean swells, be damped sufficiently rapidly. In the opposite case, a final state represented by our forced oscillations would certainly not be achieved under the actually present conditions.

In the discussion of the following section, we will thus have to keep in mind both the amplitude of the forced oscillations in equation (14) and the damping and oscillation number of the free oscillations in equation (10). In preparation, the values (14) may be written here in dimensionless form. In addition to the quantities defined in equations (9), we will use the abbreviations

$$(9''') \qquad \gamma = \frac{\omega}{\alpha_0}, \quad c = -\frac{C}{J\alpha_0^2},$$

so that γ signifies the ratio of the wave frequency to the eigenfrequency of the ship. Equation (14) is then equivalent to

$$(15) \qquad \begin{cases} \dfrac{A}{c} = \dfrac{\gamma^2 - ik\gamma - v^2}{(\gamma^2 - iK\gamma - 1)(\gamma^2 - ik\gamma - v^2) - n^2\gamma^2}, \\[2ex] \dfrac{a}{c} = \sqrt{\dfrac{J}{j}} \dfrac{in\gamma}{(\gamma^2 - iK\gamma - 1)(\gamma^2 - ik\gamma - v^2) - n^2\gamma^2}, \end{cases}$$

and the square of the amplitude ratio $\left|\frac{A}{c}\right|$ is thus

$$(16) \quad \left|\frac{A}{c}\right|^2 = \frac{(\gamma^2 - v^2)^2 + k^2\gamma^2}{[(\gamma^2 - 1)(\gamma^2 - v^2) - (kK + n^2)\gamma^2]^2 + [k\gamma(\gamma^2 - 1) + K\gamma(\gamma^2 - v^2)]^2}.$$

Parenthetically, we also verify from this formula the repeatedly emphasized fact that a fixed rotor is completely ineffective, since for $k = \infty$ the same value of A results as for $n = 0$, which we may call A_0.

The ship friction K can generally be neglected with respect to the braking friction of the stabilizer, and thus equation (16) can be replaced by

$$(17) \qquad \left|\frac{A}{c}\right|^2 = \frac{(\gamma^2 - v^2)^2 + k^2\gamma^2}{[(\gamma^2 - 1)(\gamma^2 - v^2) - n^2\gamma^2]^2 + k^2\gamma^2(\gamma^2 - 1)^2}.$$

For a ship without a stabilizer, however, the damping K may not be neglected, since, otherwise, the amplitude of the forced ship oscillation would continuously increase in the case of resonance between the ship oscillation and the waves, which is avoided only by damping.

We will have to show that for the conditions occurring in practice, a value $|A| < |A_0|$ is calculated from (16) or (17), and that, therefore, as was required at the beginning of this section, the amplitude of the forced oscillations is generally decreased by the stabilizer. We will recognize as the effective factor the quantity n, and thus the strength of the impulse of the stabilizer, while the damping k acts here in the opposite sense, and therefore against the decrease of the amplitude. We can say, on the whole, that *the inertial effect of the stabilizer is decisive for the forced oscillations, and the frictional effect of the stabilizer is decisive for the free oscillations.*

We add some references to the literature of the ship stabilizer. The differential equations for the free oscillations were first derived and discussed by H. Lorenz,*) although without consideration of the effects of braking. At almost the same time, A. Föppl**) published a detailed numerical investigation of the free oscillations, in which he explicitly pointed out the important role of the braking. The summarizing presentation that Föppl gives in the 6th volume of his *Vorlesungen über technische Mechanik* first became known to us during the proofreading of this book. We hope that our analytically more extensive presentation may suitably supplement the treatment of Föppl that depends on special knowledge and inherent intuition. A starting point for the forced oscillations is given by Malmström.***)

Schlick himself described his design, outside of a few short communications,†) in a presentation before the *Schiffbautechnische Gesellschaft*,††) in which the foundations of the theory and practical results were discussed.

We will present the latter in §6; for their understanding, we must first enter into a somewhat detailed discussion of equation (15) for the forced oscillations and equation (10) for the free oscillations.

Our gyroscopic effects (4) and (4′) in the oscillation equations are examples of "gyroscopic terms." One may imagine a mechanical system of arbitrarily many noncyclic degrees of freedom ψ, $\vartheta, \ldots$ and arbitrarily many cyclic motions φ, $\chi, \ldots$ in the form of revolving rotors. Their inertial effects are evident in the equations for ψ, $\vartheta, \ldots$ through additional "gyroscopic terms" (cf. the note to §1, IV, p. 771) that couple the motions of ψ, $\vartheta, \ldots$ with one another. For small oscillations, the gyroscopic terms are linear in the noncyclic velocities, with coefficients N, $M, \ldots$ that depend on the cyclic impulses. The

*) Physikal. Zeitschr. 5 (1903), p. 27.[302]

**) Zeitschr. d. Vereins deutscher Ing. 48 (1904), p. 478 and p. 983.[303]

***) Acta Societatis Fennicae, t. 35, 1907.[304]

†) Zeitschr. d. Vereins deutscher Ing. 50 (1906), p. 1466 and p. 1929.

††) Jahrbuch der Schiffbautechnischen Gesellschaft 1909, Nr. X. Cf. also Institution of Naval Architects, March 1904, with a theoretical appendix by A. Föppl.

general form of these terms is thus $N\psi' + M\vartheta' + \cdots$. It is now characteristic that the coefficient of ψ' vanishes in the acceleration equation for ψ, and that the coefficient of ϑ' is opposite to the coefficient of ψ' in the acceleration equation for ϑ. The coefficient schema of these terms is thus, as one may express it, "skew." It is easy to see, as on p. 801, that the energy equation is not at all influenced by the gyroscopic terms. (Cf. Natural Philosophy I, Art. 345 VII.)

The discussion of the ship stabilizer as a system of coupled oscillations has a further significance for all those problems in physics that find their expression in the consideration of coupled oscillations; we recall, for example, the Zeeman effect. The oscillation period of light waves is altered in a well-known manner if the light source is placed in a magnetic field. One may imagine a system of electrons in the light source, so that this system has a number of free oscillations, analogous to our free ship and stabilizer oscillations. The system is therefore initially characterized by a number of oscillation equations. In the magnetic field, a force perpendicular to its direction of motion, and therefore a conservative force, will be exerted on each electron, so that the total energy cannot be changed by the magnetic field. Experiments show that the force is bilinearly dependent on the velocity of the electron and the components of the magnetic field. These two conditions, however, lead to the fact (cf. above) that the magnetic field is made perceptible in the oscillation equations by the addition of "gyroscopic terms." In the place of the cyclic impulse, there now appear the components of the magnetic field strength H, and the oscillation equations thus take the form

$$m_i \frac{d^2\xi_i}{dt^2} + f_i \xi_i = \sum h_{ik} \frac{d\xi_k}{dt},$$

$$h_{ik} = -h_{ki}, \quad h_{ii} = 0,$$

where h is expressed linearly in terms of the vector H. As for the ship stabilizer, where, according to equation (10), the influence on the oscillations depends on the parameter n^2, the "displacement of the spectral lines" depends here on H^2, and is thus very small for small H. Only when the two free oscillations stand in resonance, and therefore in the case $v^2 = 1$, do we obtain from (10), with disregard of the damping,

$$\beta^2 - 1 = \pm n\beta, \quad \pm\beta = 1 \pm \frac{n}{2} + \cdots,$$

and therefore an influence on the oscillations that depends on the first power of n. Correspondingly, the magnetic displacement of the spectral lines of Zeeman were indicated only in the cases of coinciding oscillation frequencies (D-line). Similarly, we will recognize for the ship stabilizer (cf. p. 819) that coincidence of the oscillation frequencies of the ship and the rotor frame is, in a certain sense, the most favorable case; that is, the case in which the stabilizer most strongly influences, in one respect, the oscillation of the ship. For further developments, see the Encykl. d. math. Wiss. Bd. V 3, Art. 22 (Lorentz, Magnetooptische Phänomene) Nr. 31–43.

§5. Specialized discussion of the effect of the ship stabilizer.

A. Forced oscillations.

We begin the detailed discussion of the starting point formulated in the previous section by considering the influence of the stabilizer on the oscillations that are forced by the ocean swells.

If we temporarily neglect the damping K caused by friction in the water in comparison with the damping k produced by the brake, then the ratio of the amplitude A of the forced oscillation of the ship to the quantity c that measures the amplitude of the forcing waves becomes, according to equation (17),

$$\left|\frac{A}{c}\right|^2 = \frac{(\gamma^2 - v^2)^2 + k^2\gamma^2}{[(\gamma^2-1)(\gamma^2 - v^2) - n^2\gamma^2]^2 + k^2\gamma^2(\gamma^2-1)^2}. \tag{17}$$

We will consider the influence of the stabilizer on the amplitude, as on the course of the oscillation in general, as essentially dependent only on the following three dimensionless quantities:

the ratio γ of the frequency of the waves to the frequency of the free undamped oscillation of the ship;

the quantity k that measures the damping of the oscillation of the stabilizer;

the quantity n that measures the impulse imparted to the rotor.

It is of importance for the parameter v whether it is smaller or larger than 1, and thus whether the free oscillation of the stabilizer is slower or faster than the free oscillation of the ship. By adjusting the suspension of the rotor, v can easily be varied, and becomes 0 if the rotor is supported at its center of gravity.

Before entering into a numerical discussion, it is remarked that the free oscillation period of a ship varies between very wide bounds, depending on the magnitude of the oscillation. The same holds for the period of the waves, which indeed, according to general experience, takes on a known mean value for a specific sea, but varies considerably about this value and changes significantly from sea to sea. We must further imagine, as already mentioned, a superposition of many harmonic wave trains. All the more, then, will the ratio of the wave frequency to the ship frequency vary, and it is justified by practical interest to consider the parameter γ^2 as varying from 0 to ∞.

In the interest of the analytic discussion, we will begin by assuming $k^2 = 0$. This value cannot in fact be achieved, since bearing friction on the heavy rotor is always significant, even without the installation of a hydraulic brake.

An upper bound for n^2, finally, is determined only by technical difficulties.

For the graphical representation of the function $\left|\frac{A}{c}\right|^2 = f(\gamma^2, k^2, n^2)$, we will choose γ^2 as the independent variable; the quantities k^2 and n^2 will then be conceived as parameters for a system of curves. In addition, $v^2 \gtrless 1$ is to be distinguished. Our choice of the independent variable γ^2 has the advantage that we can compare with the well-known resonance curves from the theory of forced oscillations.

1. Dependence on the impulse, braking null. For $n = 0$, the case in which the stabilizer does not act on the ship, all terms in the expression (17) that depend on the coupling with the stabilizer naturally fall away, and there remains

$$\left|\frac{A_0}{c}\right|^2 = \frac{1}{(\gamma^2-1)^2}, \tag{18}$$

the simplest case of a resonance curve without damping. The amplitude corresponding to $n = 0$ is denoted by A_0; in the same manner, the amplitude for an arbitrary value of n is denoted by A_n. The amplitude A_0 increases strongly in the neighborhood of $\gamma = 1$, where the waves are in resonance with the free oscillation of the ship. The amplitude remains finite for $\gamma = 1$, however, due to the damping K that is neglected here.

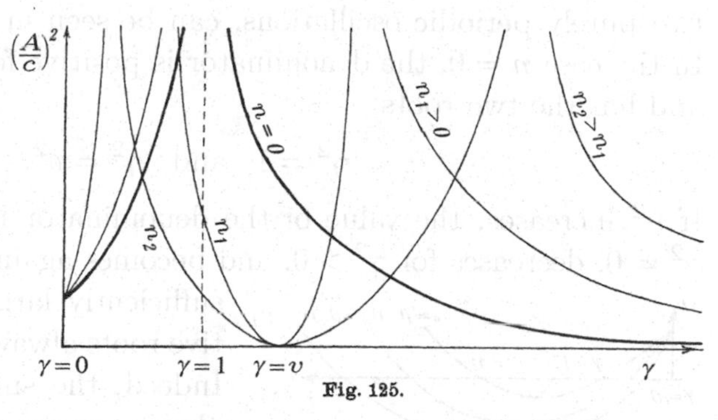

Fig. 125.

We now let the "fundamental curve" (18), drawn heavily in Figs. 125 and 127, change continuously with increasing n^2 for a fixed value of v^2. For the limiting cases $\gamma = 0$ and $\gamma = \infty$, no change can be produced by the coupling with the stabilizer. For if the period of the waves is infinitely long ($\gamma = 0$), then the equilibrium position described on p. 805 takes the place of the forced oscillation, whether the rotor spins or not. Correspondingly, the term with the factor n^2 in equation (17) is 0 for $\gamma = 0$, so that $\left|\frac{A_n}{c}\right|^2$ is always equal to 1 for $\gamma = 0$.

An infinitely rapid oscillation ($\gamma = \infty$) can excite a forced oscillation just as little when the rotor spins as when it is does not, so that $\left|\frac{A_n}{c}\right| = 0$ likewise follows from (17) for $\gamma = \infty$, independent of n. All the curves of the system corresponding to variable n thus begin from the point $\gamma = 0$, $\left|\frac{A_n}{c}\right|^2 = 1$ and asymptotically approach the γ-axis at $\gamma = \infty$.

We found in the previous section that the coupling with the stabilizer transforms the single free oscillation of the ship into two free oscillations, one of which arises from the original free oscillation of the ship and the other from the oscillation of the stabilizer that is transmitted to the ship. Resonance phenomena will occur if the period of the waves coincides with one of these two periods.

In fact, the denominator of (17), which is transformed for $k = 0$ into

$$\left|\frac{A_n}{c}\right| = \frac{(\gamma^2 - v^2)}{[(\gamma^2 - 1)(\gamma^2 - v^2) - n^2\gamma^2]}, \tag{19}$$

is the quadratic function of γ^2 that vanishes, according to equation (10) of p. 804, if the frequency of the waves coincides with one of the frequencies of the coupled system. That this denominator

$$y = (\gamma^2 - 1)(\gamma^2 - v^2) - n^2\gamma^2$$

always has two positive real roots, and that the system therefore has two purely periodic oscillations, can be seen in the following manner. In the case $n = 0$, the denominator is positive for $\gamma^2 = 0$ and $\gamma^2 = \infty$, and has the two roots

$$\gamma^2 = 1 \quad \text{and} \quad \gamma^2 = v^2.$$

If n^2 increases, the value of the denominator remains unchanged for $\gamma^2 = 0$, decreases for $\gamma^2 > 0$, and becomes again positive, however, for sufficiently large γ, so that two positive roots always remain (cf. Fig. 126). Indeed, the smaller of the two roots always moves in the direction of decreasing γ^2 and the larger in the direction of increasing γ^2, so that the smaller root will finally move to $\gamma = 0$ for $n^2 = \infty$, and the larger root to $\gamma = \infty$.

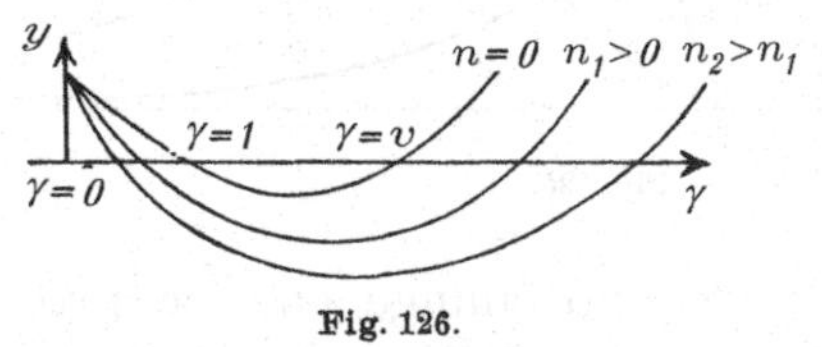

Fig. 126.

Thus as n^2 increases from 0 to positive values, the fundamental curve ($n = 0$) of Fig. 125 is transformed into a curve with two singularities, one of which lies near the previously present singularity $\gamma^2 = 1$, while the other arises in the vicinity of the position $\gamma^2 = v^2$. At the position $\gamma^2 = v^2$ itself, however, the numerator of the expression (19) retains also the value zero, so that the amplitude $|A|$ always vanishes if the waves are in resonance with the free uncoupled oscillation of the stabilizer. In the case $n^2 = 0$, the singularity and the null value of $|A|$ at $\gamma^2 = v^2$ cancel one another.

2. Distinction of the cases $v^2 \gtrless 1$. Let the singular position that arises from $\gamma = 1$ be denoted by γ_1, and that which arises from $\gamma = v$ be denoted by γ_2. We now distinguish the two cases $v^2 > 1$ (Fig. 125) and $v^2 < 1$ (Fig. 127), according as the period of the free oscillation of the stabilizer is shorter or longer than the period of the free oscillation of the ship. It may be remarked in advance that this distinction will also prove essential for the later discussion of the free oscillations.

According to the remarks concerning Fig. 126, $\gamma_1^2 < 1$ and $\gamma_2^2 > v^2$ in the first case. The amplitude therefore increases from its initial value $|A/c| = 1$ for $\gamma = 0$ and becomes infinite for the value $\gamma^2 = \gamma_1^2$. This value is always smaller, and therefore lies always farther from the original pole $\gamma^2 = 1$, as the eigenimpulse of the stabilizer is larger. As γ increases from γ_1, the amplitude decreases, and becomes 0 at the position $\gamma = v$, where the curve in Fig. 125 touches the γ-axis. The amplitude again increases for larger γ, and again becomes infinite at the position $\gamma = \gamma_2$. The value of γ_2 is always larger as n^2 is larger. For the limiting case $n = \infty$ (as already remarked), the pole γ_1 will move to 0, the pole γ_2 will move to ∞, and the amplitude will be 0 for each value of γ different from 0 and ∞.

According to the preceding, it is favorable, with respect to the forced oscillations, to suspend the rotor so that its free pendulum oscillation is as close as possible to resonance with the wave train; that is, to set $v = \gamma$. In this case, the forced oscillation of the ship would vanish altogether. The amplitude and phase of the forced oscillation of the stabilizer would then be given by (15) on p. 807 as

$$\frac{a}{c} = -\frac{i}{n \cdot v}\sqrt{\frac{J}{j}};$$

that is, the amplitude of the forced stabilizer oscillation is always smaller as the eigenimpulse of the rotor is larger, and the oscillation must be delayed in phase by a quarter-period with respect to the waves. Once excited in this phase and amplitude, the stabilizer oscillation can then exert a moment exactly opposite to the moment of the waves on the ship, so that the ship, disregarding other circumstances, remains at rest. The necessity for large n follows analytically from the fact that the amplitude of the stabilizer oscillation must become very large for small n, and thus our derivation, which indeed assumes only small oscillations, ceases to apply.

Since we must generally consider not one harmonic wave, but rather a superposition of many waves, and, in addition, the free oscillation of the stabilizer is never undamped, a precise realization of this principle of operation is certainly out the question.

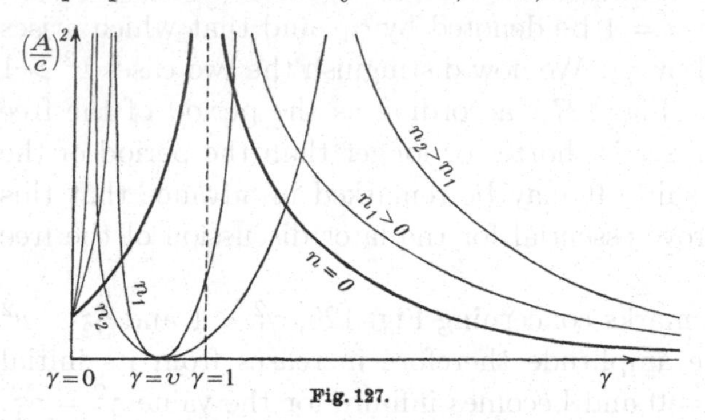

Fig. 127.

The second case $v^2 < 1$ (see Fig. 127) differs from the first only in that the amplitude first becomes infinite for the value $\gamma_2 < v$, and then becomes null for $\gamma = v < 1$, as in the first case. The amplitude then increases with increasing γ, again becomes infinite for a value $\gamma_1 > 1$, and finally decreases to zero for $\gamma = \infty$. The pole γ_2 now moves with increasing n^2 from the position $\gamma = v$ always more toward 0, and the pole γ_1 moves from 1 always more toward ∞.

3. Consequences of the effect of n. In order to obtain a measure for the favorability of the coupling with the stabilizer, we must compare the curves in Figs. 125 and 127 for finite n with the fundamental curve for $n = 0$. We first ask for the wavelengths at which the amplitude is not changed, so that $A_n = A_0$. We again begin by setting $k = 0$.

We obtain, in addition to the values $\gamma = 0$ and $\gamma = \infty$, values that are determined, according to (18) and (19), by the equation

$$2(\gamma^2 - 1)(\gamma^2 - v^2) - n^2\gamma^2 = 0.$$

There are two positive values of γ^2 that satisfy this equation; they are called γ'^2 and γ''^2. The values γ' and γ'' are therefore the abscissas of the points at which the curve for arbitrary n intersects the curve for $n = 0$. The value γ' evidently lies between 1 and γ_1, and the value γ'' between v and γ_2. We conclude from the figures that

For all waves frequencies lie between γ' *and* γ'', *the effect of the stabilizer is favorable; that is, the stabilizer decreases the amplitude of the forced oscillation of the ship. For all other waves, in contrast, the effect is unfavorable.*

It is clear, in particular, that the effect is always favorable at the position $\gamma^2 = 1$ (that is, in the neighborhood of the resonance between the waves and the free oscillation of the ship when the stabilizer is inoperative). The entire region between $\gamma^2 = 1$ and $\gamma^2 = v^2$ is certainly also favorably influenced, since $A_n = 0$ for $\gamma^2 = v^2$.

We also see that *if* $v^2 > 1$, *the favorably influenced region contains, in essence, wavelengths with* $\gamma^2 > 1$; *if* $v^2 < 1$, *it contains, in essence, wavelengths with* $\gamma^2 < 1$. With respect to the impulse strength, it follows that *by enlargement of* n^2, *and therefore the eigenimpulse, the region will be broadened on both sides, in that the positions* γ' *and* γ'' *follow the poles* γ_1 *and* γ_2.

According to the preceding, we see that the favorable effect of the stabilizer on the forced oscillations is that *it removes, for an appropriate suspension, the frequencies of the eigenoscillations of the ship as far as possible from the frequency of the waves, therefore eliminating the detrimental resonance effects.*

4. Dependence of the braking k. We now consider the previously neglected damping k. We can easily discuss its effect if we correspondingly modify the previously obtained figures.

We remark, to this end, that the complete expression (17) for the value of the amplitude can be written, if we set $y = |A/c|^2$ and imagine that the denominator is multiplied out, as

$$f(y, \gamma) + k^2 g(y, \gamma) = 0. \tag{18c}$$

Here $f(y, \gamma) = 0$ signifies one of the previously drawn curves that corresponds to a particular value of n in the case $k = 0$. For $k = \infty$, on the other hand, we have $g(y, \gamma) = 0$; this curve must coincide with our "fundamental curve" for $n = 0$, as one easily confirms from (17), since indeed a stabilizer with infinitely strong braking is just as ineffective as one that does not spin. The intersection points of the curves $f = 0$ and $g = 0$ were just designated as γ' and γ''; according to (18c), the curve for an arbitrary k must also pass through these points.

At these two positions γ' and γ'', which depend on the values of n^2 and v^2, the damping has no influence on the amplitude of the forced oscillation. For other values of γ^2, it is to be noted that $|A/c|^2$, as a linear function of k^2, always changes with increasing k^2 in the same sense, and neither vanishes nor becomes infinite at any finite positive value of k^2. We therefore obtain the bundle of curves that corresponds to fixed n^2 and v^2 and variable k^2 if we transform the corresponding curve in Fig. 125 or Fig. 127 continuously into the fundamental curve $k = \infty$ (cf. Fig. 128). The ordinate therefore creases wherever it was less than that of the fundamental curve (18), and decreases wherever it was greater. *The effect of the stabilizer on the amplitude of the forced oscillations is everywhere decreased by the braking k.* If the effect without damping is favorable, it will become more unfavorable with damping; if it is unfavorable, it will become more favorable. Thus we can now rightfully formulate the result that *the favorable effect of the stabilizer on the forced oscillations is only that it distances the frequency of the free ship oscillation, for an appropriate suspension and an appropriate impulse strength, as much as possible from the frequency of the waves.*

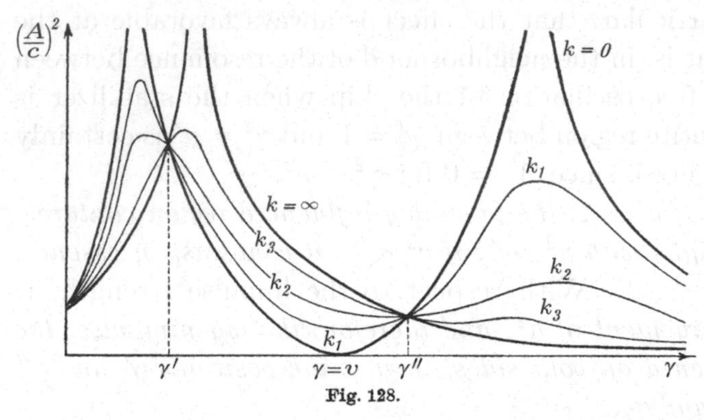

Fig. 128.

We have thus justified the first part of the claim made at the end of the preceding section (p. 808), that *the inertial effects of the stabilizer are the decisive factors for the forced oscillations*, while the damping only decreases the effect here. We emphasize again, however, that sufficient damping of the free oscillations by the brake is a precondition for the pure occurrence of the forced oscillations.

B. Free oscillations.

1. Primary oscillation and secondary oscillation. We must ask, in the following, for the period and the damping of the free oscillations of the ship as they are affected by the stabilizer; that is, we must discuss the roots of the equation

$$(\beta^2 - 1)(\beta^2 - ik\beta - v^2) - n^2\beta^2 = 0 \tag{10}$$

on p. 804 for variable n^2, v^2, and k. As we remarked in the preceding section (p. 802), the single free oscillation of the ship is replaced by two oscillations through the coupling to the stabilizer. One oscillation (the primary oscillation) arises from the original oscillation of the ship, and the other (the secondary oscillation) from the free pendulum oscillations of the stabilizer and the coupling to the ship by means of the eigenimpulse of the rotor.

We could solve the equation of the 4th degree for β algebraically; its conjugate pairs of roots would then determine the oscillations. For a qualitative discussion, however, little would be gained in this manner. It could hardly be seen from the complicated form of the solution, in particular, how the two oscillations arise continuously from the original oscillations of the ship and the stabilizer as n^2 increases. It is better, for this purpose, to use "continuity methods."

For the undamped case $k = K = 0$, we have already seen in the section on the forced oscillations (Fig. 126, p. 812) that the faster of the primary and secondary oscillations always becomes still faster with increasing n^2, and the slower still slower. The period of the ship oscillation will therefore be lengthened if the period of the stabilizer oscillation is shorter than that of the ship oscillation, and will be shortened if the period of the stabilizer oscillation is longer. This corresponds to the motion of the poles in the resonance curves of Figs. 125 and 127. For infinitely large n^2, finally, the period of one oscillation will be 0, and that of the other ∞.

If we now consider damping, then we must represent the roots of equation (10) not as points on the real axis as in Fig. 126, but rather as points in a complex plane $x + iy$. The real part x then signifies the frequency and the imaginary part iy the damping, so that $2\pi y/x$ signifies the logarithmic decrement, as long as the oscillation is not damped aperiodically. We can naturally no longer apply the simple argument that led to the course of the roots in Fig. 126.

It will be shown in the following discussion that none of the roots of (10) can now be purely real, as they were in the undamped case, even if $K = 0$, except for $n^2 = 0$ or $n^2 = \infty$. Both oscillations are therefore damped. Pure imaginary roots, and therefore aperiodically damped oscillations, are indeed possible. Negative imaginary roots, however, which correspond to an aperiodic increase of the deflection of the ship or the stabilizer, are excluded by the form of equation (10), since a negative imaginary root would make the left-hand side of (10) a sum of only positive terms. We will show in our more detailed discussion that there are also no roots with negative imaginary parts, and therefore no increasing oscillations (cf. p. 824); this can, moreover, be foreseen without proof, since such solutions would be incompatible with the strictly negative dissipation function in the energy principle of p. 801.

In order to distinguish the roots corresponding to the two oscillations, the pair of roots corresponding to the *primary oscillation* will be denoted by ζ_1, ζ_2, and the pair corresponding to the *secondary oscillation* by η_1, η_2. For $n^2 = 0$,

$$\zeta_1, \zeta_2 = \pm 1,$$

$$\eta_1, \eta_2 = \frac{ik}{2} \pm \sqrt{v^2 - \frac{k^2}{4}}.$$

For variable n^2, the four roots describe curves in the complex plane that we must investigate. As long as the oscillations are periodic, the curves for ζ_1 and ζ_2 and for η_1 and η_2 will be symmetric about the imaginary axis;*) for aperiodic behavior, the curves come together onto the imaginary axis.

2. Approximations for small and large impulse. We first investigate how the roots of (10) change if n^2 increases from 0 to a small positive number, where $K = 0$ is again assumed for the sake of greater transparency. We remark, however, that the following method may also be applied in just the same manner to the case $K > 0$, and the results are then changed only inessentially.

Equation (10) can be written as an equation for the pair ζ_1, ζ_2 as

$$\zeta^2 - 1 = \frac{n^2}{\zeta^2 - ik\zeta - v^2}.$$

We seek the root ζ_1 that arises from the root $\zeta_1 = +1$, and thus set the first approximation $\zeta = +1$ on the right-hand side. We thus obtain the second approximation

$$\begin{aligned}\zeta_1 &= +\sqrt{1 + \frac{n^2}{1 - v^2 - ik}} \\ &= 1 + \frac{1}{2}\frac{n^2(1 - v^2)}{(1 - v^2)^2 + k^2} + \frac{i}{2}\frac{n^2 k}{(1 - v^2)^2 + k^2},\end{aligned}$$

where the square root is expanded in n^2, and only the first term is considered. In the same manner,

$$\zeta_2 = -1 - \frac{1}{2}\frac{n^2(1 - v^2)}{(1 - v^2)^2 + k^2} + \frac{i}{2}\frac{n^2 k}{(1 - v^2)^2 + k^2}.$$

There follows:

*) A root pair of the form $a + ib$, $-a + ib$ is called "conjugate real," in analogy to the term "conjugate imaginary."

If the oscillation of the ship is undamped for $n = 0$, it will be damped as n^2 increases.

The frequency of the primary oscillation decreases or increases, according as the frequency of the stabilizer oscillation is greater than or less than ($v^2 \gtrless 1$) that of the ship oscillation.

We also confirm once more that the effect of the stabilizer vanishes if the damping k becomes infinitely large. On the other hand, the damping increase $\frac{n^2}{2}\frac{k}{(1-v^2)^2+k^2}$ naturally also vanishes if $k=0$. Between $k=0$ and $k=\infty$ the damping must therefore have a maximum; differentiation of the given damping increase shows that the maximum occurs for the value

$$k_{\max}^2 = (1-v^2)^2.$$

While the damping of the primary oscillation increases, the damping of the secondary oscillation decreases from its original value $ik/2$ (cf. p. 818), since the sum of the damping constants (that is, the sum of the four roots or, the coefficient of $-i\beta^3$ in (10)) remains constant. This conclusion is naturally to be modified for the case of an aperiodically damped secondary oscillation, which is usually present in practice.

We ask, finally, how the period of the pendulum oscillation of the stabilizer, and therefore v^2, is to be chosen in order that the damping of the primary oscillation should increase most rapidly with n^2 for fixed k. Excluding the case $k=0$,

$$\frac{n^2 k}{(1-v^2)^2+k^2}$$

will obviously be maximum when $v^2 = 1$, and thus when the free pendulum oscillation of the stabilizer and the free oscillation of the ship are in resonance (cf. the endnote to §4, p. 809). As long as the secondary oscillation does not come into consideration, the suspension of the stabilizer is most favorable, at least for sufficiently small values of n^2, when $v^2 = 1$.

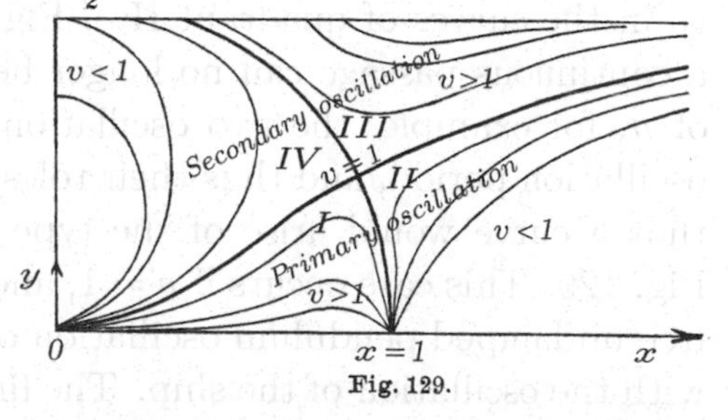

Fig. 129.

According to the preceding results for small n, the root ζ_1 moves from its initial value 1 (Fig. 129) in the direction of increasing y and in the direction of increasing or decreasing x, according as $v^2 \lessgtr 1$. We now seek the end of our curves; that is, the behavior of the roots for the limiting value of the eigenimpulse $n^2 = \infty$.

For $n^2 = \infty$, equation (10) is satisfied by one pair of roots $\beta^2 = 0$; we initially leave undecided whether this pair of roots corresponds to the primary oscillation ζ or the secondary oscillation η. As a second approximation one then obtains

$$\beta^2 = \frac{v^2}{n^2}.$$

A second possible pair of roots is $\beta^2 = \infty$. The second approximation in this case is

$$\beta^2 - ik\beta - (1 + v^2 + n^2) = 0,$$

and therefore

$$\beta = \frac{ik}{2} \pm \sqrt{-\frac{k^2}{4} + (1 + v^2 + n^2)}.$$

As in the undamped case, we thus have, in the first approximation for infinitely large eigenimpulse, the free frequencies 0 and ∞. The first oscillation is also undamped in the second approximation; the second oscillation has the damping constant $k/2$, as in the case $n = 0$.

3. Different possibilities for the passage from small to large impulse. According to the circumstances, a first possibility is that the original oscillation of the ship is transformed continuously with increasing n^2 into the undamped oscillation with frequency 0. The damping would then have a maximum, for a certain value of n, that could not be surpassed by enlargement of the impulse (cf. the curves in quadrant I of Fig. 129; in this figure it is to be imagined that k is constant, n varies on an individual curve from 0 to ∞, and v changes from curve to curve as the parameter of the set). It is also possible, however, that the oscillation of the ship is transformed continuously into the infinitely rapid oscillation, which would then assume the complete damping of the stabilizer oscillation, as in the curves of quadrant II of Fig. 129. Finally, it is possible that a continuous passage can no longer be spoken of, if for a certain value of n, for example, the two oscillations coincide in both damping and oscillation period, and thus their roles could be partly interchanged, so that a curve would arise of the type that separates the quadrants in Fig. 129. This case occurs if $v = 1$, and therefore if the frequency of the free undamped pendulum oscillation of the stabilizer initially coincides with the oscillation of the ship. The first possibility is realized by $v > 1$ and the second by $v < 1$, as indicated in the figure.

The curves in quadrants III and IV of Fig. 129 represent the root values corresponding to the secondary oscillation, and indeed in such a manner that each curve of quadrants I and III corresponds to a curve of quadrants II and IV with equal value v^2. Since k, as we have emphasized, is the same for all curves of the figure, all curves for the secondary oscillation begin for $n = 0$ at points on the horizontal line $y = k/2$, and indeed at positions that depend on v; only for small v, when the original pendulum oscillation of the stabilizer is aperiodic, does the curve begin from the imaginary axis. For the meaning of v in the aperiodic case, where we can no longer speak of a ratio of oscillation numbers, cf. the note to p. 804.

Since the achievable degree of damping for the oscillation of the ship, and thus the favorability of the entire system, depends directly on these different possibilities of the passage from small to large n, we wish to investigate this passage in detail, and thus give a rigorous proof of the previously made claims. With respect to the choice of the impulse strength, we can already establish a result: if the course of the roots follows a curve in quadrant I, then the damping of the oscillation of the ship achieves a maximum for a certain value of n^2, and it would be undesirable, with respect to the damping, to increase the impulse above this value. If, in contrast, the roots follow a curve in the second quadrant, then the oscillation of the ship can assume the full damping k of the stabilizer pendulum for a sufficiently elevated impulse strength, and it is thus to be expected that the damping initially increases here much more rapidly with increasing n^2 than in the first case. A glance at Fig. 129 shows, however, that the damping increases only slowly for large n^2, and that the frequency of the oscillation of the ship also increases, which is undesirable. Thus the impulse will here too have an upper bound, above which a further increase will be of little value with respect to the damping and undesirable with respect to the frequency. Numerical calculation confirms, in conformity with experience, that a favorable mean value of n^2 exists.

4. Quantitative determination of the roots for an arbitrary impulse. Differential law for the motion of the roots. We now investigate these phenomena by extending the approximation methods that were applied for $n^2 = 0$ and $n^2 = \infty$.

Written out, equation (10) is

$$\beta^4 - i(k+K)\beta^3 - (1 + v^2 + kK + n^2)\beta^2 + i(Kv^2 + k)\beta + v^2 = 0.$$

The coefficients of this equation give conditions for its roots, which we immediately write in the form

$$(20)\quad \begin{cases} (\zeta_1+\zeta_2)+(\eta_1+\eta_2)=i(k+K) \\ \zeta_1\zeta_2+\eta_1\eta_2+(\zeta_1+\zeta_2)(\eta_1+\eta_2)=-(1+v^2+kK+n^2) \\ \zeta_1\zeta_2\cdot(\eta_1+\eta_2)+\eta_1\eta_2\cdot(\zeta_1+\zeta_2)=-i(Kv^2+k) \\ \zeta_1\zeta_2\cdot\eta_1\eta_2=v^2. \end{cases}$$

With consideration of the reality conditions, we set for the products and sums

$$(21)\quad \begin{cases} \zeta_1\zeta_2=-p; & \eta_1\eta_2=-\pi; \\ \zeta_1+\zeta_2=+is; & \eta_1+\eta_2=+i\sigma. \end{cases}$$

Here s and σ signify the respective damping factors of the two oscillations, and $\sqrt{p}$ and $\sqrt{\pi}$ the instantaneous "undamped" frequencies that would be present for $s=0$ and $\sigma=0$.

Since n^2 is present only in the coefficient of β^2, it follows by differentiation of the above equations with respect to n^2 that

$$\begin{aligned} s'+\sigma' &= 0 \\ \sigma s'+s\sigma'+p'+\pi' &= 1 \\ \pi s'+p\sigma'+\sigma p'+s\pi' &= 0 \\ \pi p'+p\pi' &= 0, \end{aligned}$$

where the prime denotes differentiation with respect to n^2.

Finally, the solution of these linear equations for the differential quotients gives, if Δ denotes the determinant of the system, the formulas

$$(22)\quad \begin{cases} \Delta\cdot s' = -p\sigma+\pi s \\ \Delta\cdot\sigma' = +p\sigma-\pi s \\ \Delta\cdot p' = p(\pi-p) \\ \Delta\cdot\pi' = \pi(p-\pi). \end{cases}$$

The determinant Δ is

$$(23)\quad \Delta=\begin{vmatrix} 1 & 1 & 0 & 0 \\ \sigma & s & 1 & 1 \\ \pi & p & \sigma & s \\ 0 & 0 & \pi & p \end{vmatrix}$$

$$=-(p-\pi)^2+(s-\sigma)(p\sigma-\pi s),$$

or, if we substitute from (21),

$$\begin{aligned} \Delta &= -(\zeta_1\zeta_2-\eta_1\eta_2)^2+[(\zeta_1+\zeta_2)-(\eta_1+\eta_2)][\zeta_1\zeta_2(\eta_1+\eta_2)-\eta_1\eta_2(\zeta_1+\zeta_2)] \\ &= -(\zeta_1-\eta_1)(\zeta_1-\eta_2)(\zeta_2-\eta_1)(\zeta_2-\eta_2). \end{aligned}$$

The determinant therefore vanishes only if one root of the pair ζ_1,ζ_2 coincides with one of the pair η_1,η_2. As long as this does not occur, the determinant always retains its original sign as n^2 changes. Only in the exceptional case can the quantities p, π, s, σ change discontinuously with n^2; otherwise, their change occurs with a continuous sense of progression.

Formulas (22) now permit us to discuss, with consideration of this remark, the curves for the roots as n^2 varies. For their actual calculation, we begin with n^2 equal to 0, let it increase by a small value $\delta(n^2)$, and consider the changes of p, π, s, σ in this interval as linear, according to the example of "mechanical quadrature." In this manner, in any case, the roots are obtained for a continued series of values of the impulse n^2 much more rapidly and transparently than by the direct numerical solution of the equation of the fourth degree.

Equations (22) first yield only the real quantities p, π, s, σ. The roots themselves are obtained by solving the quadratic equations

$$\zeta^2 - is\zeta - p = 0$$
$$\eta^2 - i\sigma\eta - \pi = 0,$$

so that

$$\text{(24)} \qquad \begin{aligned} \zeta_1, \zeta_2 &= +i\frac{s}{2} \pm \sqrt{-\frac{s^2}{4} + p} \\ \eta_1, \eta_2 &= +i\frac{\sigma}{2} \pm \sqrt{-\frac{\sigma^2}{4} + \pi}. \end{aligned}$$

For the initial value $n^2 = 0$, the quantities (21) have the values

$$\text{(25)} \qquad p = 1; \quad \pi = v^2; \quad s = K; \quad \sigma = k,$$

so that the determinant Δ is approximately, because of the smallness of K,

$$\text{(25a)} \qquad \Delta = -(1 - v^2)^2 - k^2.$$

Here p and π are positive, and also remain positive for increasing n^2, since, according to (20) and (21), their product has the invariable value v^2. The determinant Δ is, according to (25a), certainly negative for $n^2 = 0$, and therefore remains negative as long as one root of the ζ-pair does not coincide with one of the η-pair. We will first assume, as in Fig. 129, that the coupling does not suffice to make the oscillation of the ship aperiodic. Then at least one pair of roots is a conjugate pair, and a coincidence can thus occur only if the two pairs coincide completely, so that $p = \pi$ and $s = \sigma$. If this does not occur, then the determinant Δ always remains negative.

We remark, finally, that it is easily shown from equations (22) that the quantities s and σ never become negative, so that the two oscillations cannot increase temporally, and thus the method of small oscillations actually remains applicable. For it is not possible for any finite value of n^2 that p or π will become 0 or ∞, since equation (10) can have vanishing or infinitely large roots only for infinitely large n^2. Now s and σ are positive for small n^2, and it follows from equations (22), because of the negative value of Δ, that s and σ always increase if they are initially very small. Thus it is not possible, for any finite positive value of n^2, that one of the quantities s or σ vanishes and later becomes negative. It is tacitly assumed in this proof that Δ does not vanish; this exceptional case, which has already been mentioned, occurs for $v^2 = 1$, and will be treated specifically in No. 6.

5. General course of the roots for $v^2 \gtrless 1$. After these preparatory remarks, the course of the curves for the roots (Fig. 129) is easy to discuss. It is to be noticed that in this and the following figures the abscissa x does not represent the quantity p or π, but rather the frequency $\sqrt{p^2 - \left(\frac{s}{2}\right)^2}$ or $\sqrt{\pi^2 - \left(\frac{\sigma}{2}\right)^2}$, which has, however, behavior similar to that of p or π itself.*) It is obvious from equations (22) that the initial values $\pi - p \gtreqless 0$ should be distinguished; we must therefore consider, according to (25), the cases

$$v^2 \gtreqless 1.$$

To begin, let

$$v^2 > 1,$$

so the original pendulum oscillation of the stabilizer is more rapid than the original oscillation of the ship.

It then follows from formulas (22), as long as Δ is negative, that p' is negative, π' is positive, p decreases from the value 1, and π increases from the initial value $v^2 > 1$, so that p cannot equal π. Therefore the determinant Δ cannot vanish, and p constantly decreases further and π increases further. The primary oscillation must finally (according to Fig. 129) be transformed for large n^2 into the undamped oscillation with frequency 0, and the secondary oscillation into the damped oscillation with frequency ∞. The case $v^2 > 1$ therefore corresponds,

*) Corresponding to the case that is present in practice, it is assumed in Figs. 130 and 131 that the free oscillation of the stabilizer is aperiodically damped, so that the two roots η lie on the imaginary axis for small n^2. Only in Fig. 129, for the sake of greater clarity, are the periodic cases of the secondary oscillation also shown.

as was stated previously without proof, to the first of the possibilities discussed on p. 820, No. 3, and the corresponding curves lie in quadrants I and III. *The damping of the primary oscillation must achieve a maximum for some value of n^2, and then decrease.* The formulas (22) confirm this behavior, since the term in the equation for s' with the constantly increasing factor π dominates for sufficiently large n^2, so that, because of the negative value of Δ, s' must finally become again negative. The damping σ of the secondary oscillation, which is represented by points in quadrant III, achieves, at the same time, a minimum value, and then increases again to its original value $\sigma = k$.

In the case

$$v^2 < 1,$$

when the pendulum oscillation of the stabilizer is slower than the oscillation of the ship, it follows from (22), analogously to the case $v^2 > 1$, that p constantly increases from the value 1, and that π, in contrast, beginning from $v^2 < 1$, constantly decreases. The primary oscillation is now transformed for large n^2, according to Fig. 129, into the oscillation with infinitely large frequency that has the full damping k, and the secondary oscillation approaches the undamped oscillation with frequency 0. The case $v^2 < 1$ therefore corresponds, as we claimed above, to the second of the possibilities named on p. 820, and the comments made there apply to this arrangement. Formulas (22) confirm that the term with the constantly increasing quantity p now always dominates in the expression for s', so that s always increases.

In reality, only a small part of the curves in Fig. 129 actually comes into consideration, since the magnitude of n^2 is indeed bounded by technical difficulties. In any case, however, a greater damping of the primary oscillation will be achieved, according to the discussion above, if the construction of the stabilizer is chosen so that it oscillates more slowly than the ship when the rotor does not spin. The "undamped" frequency of the primary oscillation (p) will indeed be increased at the same time, which must finally lead to an undesirable increase of the actual frequency $x = \sqrt{p^2 - \frac{s^2}{4}}$, while for moderate values of n the actual frequency can still decrease because of the increasing damping. On the other hand, it is advantageous that the secondary oscillation, always chosen in practice to be aperiodic, remains aperiodic for greater values of the impulse, while in the case $v^2 > 1$ the secondary oscillation will become periodic much sooner. The secondary oscillation will also become periodic for large n^2 in the case $v^2 < 1$, but with a very long period. Numerical calculations confirm these conclusions, *which show that it is favorable to let the stabilizer oscillate as slowly as possible, and in no case faster than the ship.* The stabilizer must still possess, however, a gravitational moment that is sufficient to return it automatically to its vertical equilibrium, in spite of the friction, after the cessation of the rolling motion.

6. Special treatment of the case $v^2 = 1$.*) We have still to treat of the resonant case $v^2 = 1$, in which the initial values of p and π are certainly $p = \pi = 1$. It follows from equations (22), however, that the values

$$p = \pi = 1 \tag{26a}$$

initially remain valid for increasing n.

For Δ cannot vanish as long as the two root pairs are not completely identical, and therefore $s = \sigma$. Equations (22) simplify with $p = \pi = 1$ to

$$s' = \frac{1}{\sigma - s}; \quad \sigma' = \frac{1}{s - \sigma}.$$

Thus it follows that s increases and σ decreases until $s = \sigma$. By subtraction of these equations there follows

$$\frac{d(s-\sigma)}{d(n^2)} = \frac{-2}{s-\sigma},$$

and therefore

$$(s-\sigma)^2 = -4n^2 + \text{const.},$$

where the constant is given by (25) for the case $n = 0$ as k^2, if $K = 0$ is assumed.

It follows that

$$s - \sigma = -\sqrt{-4n^2 + k^2}.$$

Further, according to equation (10),

$$s + \sigma = k.$$

Therefore

$$\begin{cases} s = \frac{1}{2}(k - \sqrt{k^2 - 4n^2}\,) \\ \sigma = \frac{1}{2}(k + \sqrt{k^2 - 4n^2}\,). \end{cases} \tag{26b}$$

Thus $s = \sigma$ for the value $n^2 = \dfrac{k^2}{4}$, at which, because of (26a), the two root pairs become identical.

In the case $v^2 = 1$, the equation of the fourth degree (10) is solvable by mere square roots.

*) The treatment of the problem by A. Föppl cited on p. 808 is restricted to this mathematically somewhat easier special case.

Its complete solution, according to (26a), (26b) and the schema (24) of p. 823, is

$$(27)\quad \begin{cases} \zeta_1, \zeta_2 = \frac{i}{4}\left(k - \sqrt{k^2 - 4n^2}\right) \pm \sqrt{1 - \frac{1}{16}\left(k - \sqrt{k^2 - 4n^2}\right)^2} \\ \eta_1, \eta_2 = \frac{i}{4}\left(k + \sqrt{k^2 - 4n^2}\right) \pm \sqrt{1 - \frac{1}{16}\left(k + \sqrt{k^2 - 4n^2}\right)^2}. \end{cases}$$

For further increase of n^2, our values s and σ would themselves become imaginary. Two other pairs of roots then become conjugate pairs, and these would finally go over, as in the cases $v^2 \gtrless 1$, into the limiting values 0 and ∞, as is also to be seen from equations (27). A continuous passage of the ship oscillation or the stabilizer oscillation into one of the finally achieved oscillations can then, however, no longer be spoken of.

According to the preceding, the curve for the case $v^2 = 1$ separates the drawing plane, as in Fig. 129, into four quadrants in which the curves for the values $v^2 \neq 1$ must be ordered, and indeed without any two curves intersecting. For since the left-hand side of equation (10) contains the parameters v^2 and n^2 linearly, these parameters are determined uniquely by the specification of a pair of roots, so that, in general, only one curve can pass through a point of the plane.

A general overview of the behavior of the roots for variable n^2 may thus be acquired. The latterly treated case $v^2 = 1$ was earlier (p. 819) found to be the arrangement in which, at least for small impulse, the strongest damping of the primary oscillation is achieved for a given impulse of the stabilizer. The value $v^2 = 1$ remains the most favorable case in this respect for greater impulse values as well. For with the prime now denoting differentiation with respect to v^2 for fixed n^2 and k (not, as previously, with respect to n^2 for fixed v^2 and k), there follow from equation (20) and (21)

$$\begin{aligned} s' + \sigma' &= 0 \\ \pi s' + p\sigma' + \sigma p' + s\pi' &= 0 \\ \pi p' + p\pi' &= 1 \\ \sigma s' + s\sigma' + p' + \pi' &= 1. \end{aligned}$$

For the case of resonance, in which, according to (26a), $p = \pi = 1$, at least as long as $n^2 < k^2/4$, there follow

$$\begin{aligned} \sigma s' + s\sigma' &= 0, \\ s' + \sigma' &= 0, \end{aligned}$$

and therefore

$$s' = 0; \ \sigma' = 0.$$

For a prescribed value of n^2, both dampings thus have extremal values in the case of resonance; specifically, as may be shown in a well-known manner and was already found for small n^2, the primary oscillation has a maximum, which is desirable, and the secondary oscillation has a minimum, which is undesirable, and all the more undesirable in the present case of resonance, since the amplitude of the secondary oscillation must also become relatively large.

If it were a matter only of the damping of the primary oscillation, the choice $v^2 = 1$ would therefore be the most favorable arrangement. It is to be considered, however, that the corresponding secondary oscillation quickly loses damping and can also become perceptible because of its relatively strong amplitude, while for smaller values of v^2 the secondary oscillation remains aperiodic or slowly periodic much longer because of its small frequency, and certainly does not become damaging.

7. Possibility of aperiodic decay of the primary oscillation. We have restricted ourselves so far to the case in which the oscillation of the ship does not become aperiodically damped. As our following numerical calculations will show, however, this restriction is not always necessary under the conditions that occur in practice, at least not for the arrangement $v^2 \leq 1$.

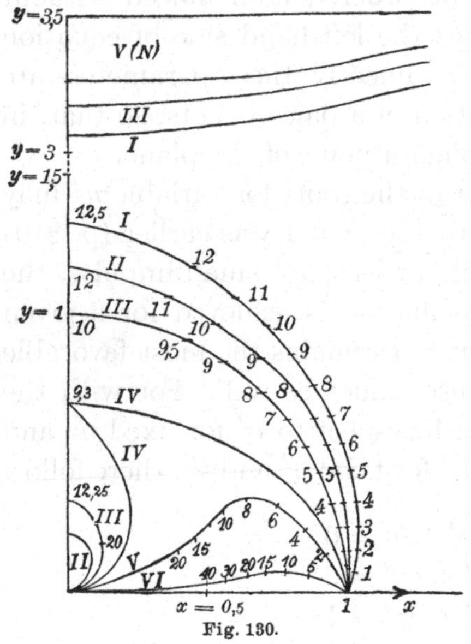

Fig. 130.

Curve	I	II	III	IV	V	VI
$v^2 =$	0	0,25	1	1,5	4	20

(N) = Secondary oscillation.

The numbers denote the values of n^2. The points are calculated according to the method of No. 4 for k=7.

The form of the curves, which may be contrasted with those in Fig. 129, is represented in Fig. 130. As in Fig. 129, the value of k is constant, but is to be imagined as larger than the value there. The parameter n varies on an individual curve, and the parameter v varies from curve to curve. For not too large values of v (cf. I to IV in the figure), the curves rise from their common starting point on the x-axis until they reach the y-axis. The primary oscillation then becomes aperiodic. If both oscillations are aperiodic, and therefore the four roots of (10) are purely imaginary, then a distinction of the primary and secondary oscillations can no longer be spoken of, since the separation of the roots into two pairs ceases to occur. For very large n^2, in contrast, the oscillations must again become periodic, since for $n^2 = \infty$ there always remain two periodic oscillations with the frequencies 0 and ∞. The damping of the fast oscillation in this case does not exceed the value k, and the damping of the slow oscillation approaches zero. Correspondingly, we see developing from curve III in Fig. 130, for example, an upper branch and a lower branch lying in the vicinity of the origin, which are to be imagined as bound to the original branch of curve III by parts of the imaginary y-axis.

It is in conformity with our earlier remarks that aperiodicity occurs only for intermediate values of n^2, and that there is always an upper bound above which it is purposeless to increase the impulse, and in part even damaging. The domain of aperiodicity permits us to discuss somewhat quantitatively the bounds in which the strength of the impulse is to be maintained, if possible.

8. F a v o r a b l e b o u n d s f o r t h e s t r e n g t h o f t h e i m p u l s e a n d t h e s t r e n g t h o f t h e b r a k i n g. While k is constant in Figs. 129 and 130 and n and v are variable, v is constant in Fig. 131; n varies on an individual curve and k varies from curve to curve. And indeed we have chosen $v^2 = 1$, in which case the curves may be most easily computed numerically. The numbers attached to the marks signify the values of n^2; the calculation of the points is made according to the formulas of No. 6. It is inessential that the curves here, similar to the curve for $v^2 = 1$ in Fig. 129, have a discontinuity in direction when the primary and secondary oscillations coincide. In general, the same image would result for a different value of v^2, only the curves

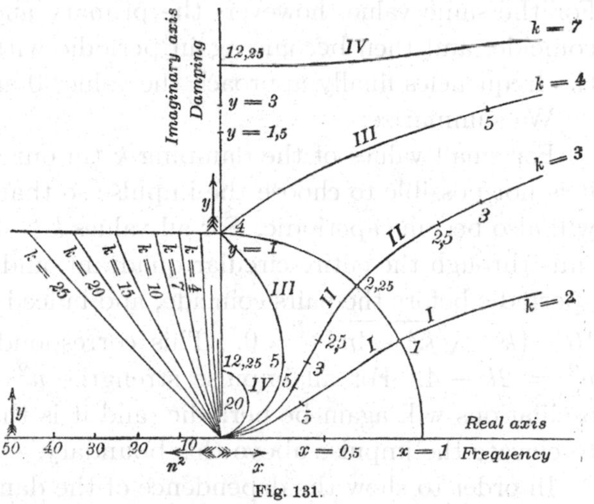

Fig. 131.

would remain continuous in the vicinity of the kink of our figure, as in Fig. 129 for $v^2 \neq 1$. It is likewise inessential, and due only to the special choice $v^2 = 1$, that the roots initially move on the same circle for all values of k (since here the product of the two conjugate real roots is $\zeta_1\zeta_2 = (x+iy)(-x+iy) = -1$).

Curve I of Fig. 131 gives the course of the roots for the small value $k = 2$. Here the root η corresponding to the secondary oscillation begins for $n^2 = 0$ from the value $+i$; it stands on the boundary between periodicity and aperiodicity, and moves with increasing n^2 on the circle $x^2+y^2 = 1$. The roots ζ corresponding to the primary oscillation begin from the points ± 1 (only the root with positive real part is shown in the figure) and move upward on the named circle. According to equations (26) of p. 826, the two pairs of roots coincide when

$$n^2 = \frac{k^2}{4} = 1.$$

For greater values of n^2, s and σ become imaginary; it is then evident from equations (27) that none of the 4 roots can be purely imaginary (except in the limit $n = \infty$), and therefore the oscillations cannot be aperiodic. The same holds for curve II, $k = 3$, only here the secondary oscillation is initially aperiodic; it becomes periodic, however, for $n^2 = 2$. For $n^2 = 2{,}25$, the two pairs coincide and then both remain periodic.

The value $k = 4$ is chosen for curve III. Here the damping s increases until it reaches the value 2, and the primary oscillation reaches the boundary between periodicity and aperiodicity at the value $n^2 = 4$. For the same value, however, the primary and secondary oscillations coincide, and then become again periodic with increasing n^2, so that the frequencies finally approach the values 0 and ∞.

We summarize:

For small values of the damping k (in our case $v^2 = 1$, for $k \leq 4$), it is not possible to choose the impulse so that the primary oscillation will also become aperiodic. For all values $k > 4$, in contrast, the root ζ runs through the entire circular quadrant, and the oscillation becomes aperiodic before the pairs coincide, and indeed according to (27), when $16 - \left(k - \sqrt{k^2 - 4n^2}\right)^2 = 0$. This corresponds to the positive value $n^2 = 2k - 4$. For all impulse strengths $n^2 > k^2/4$, however, both oscillations will again be periodic, and it is thus certainly purposeless to elevate the impulse above this boundary.

In order to show the dependence of the damping on the impulse n^2 for the points of the circular quadrant, we have plotted to the left of the imaginary axis the curves

$$\frac{s}{2} = f(n^2),$$

and thus the imaginary part of ζ as a function of n^2, and indeed for different values of the damping k. The numbers on the horizontal axis denote the values of n^2. It is seen from the curves that for *sufficiently large k (in the case $v^2 = 1$, for $k > 4$), the oscillation can always become aperiodic, but the required impulse strength is always larger as the damping k is larger.*

It may be remarked parenthetically here that in the case of the stabilizer designed for the "Silvana," n^2 cannot be increased significantly above 20. The curves then show that k may not be larger than 10 if aperiodicity of the primary oscillation is still to be possible; the corresponding value of the braking strength is $w = 1000$ mkgsec.

The relations described in the preceding for $v^2 = 1$ also give a sufficient image for the corresponding relations at other values of v^2.*) The effect of the brake, as given in the preceding, may thus be generalized in the following manner:

The damping factor s of the primary oscillation increases rapidly with n^2 for small k, but increases more slowly as k becomes larger (cf. the left-hand part of Fig. 131), *and can never exceed the value of k itself. If it reaches the vicinity of this value, it increases only slowly, and approaches the limiting value k asymptotically* (cf. the right-hand part of Fig. 131). *Thus greater values of s can generally be achieved for greater values of k, but only under the assumption that the impulse can be increased arbitrarily.*

Because of the realistic limit for the strength of the impulse, the question in practice is to achieve the most favorable effect possible by the choice of k for a fixed value of n^2. According to the previous deliberations, such a most favorable value of the braking always exists. If we assume, namely, a large value of k, then the point that corresponds to the fixed value of n^2 lies on the initial steep ascent of the relevant curve (in the case of Fig. 131, on the circular arc), and indeed always lies higher as k is smaller; if we choose, in contrast, a small value of k, then the point lies on one of the associated branches of the curve with a smaller ascent (in the case of Fig. 131, on one of the arcs that branch from the circle to the right), and therefore (cf. Fig. 131) always lies

*) Since we concluded that the case $v^2 > 1$ is unfavorable, $v^2 = 1$ forms the upper and $v^2 = 0$ the lower bound for v. In the latter case, the analogous results may be derived analytically by the formation of the discriminant of equation (10). In this case, for example, the most favorable braking strength for a given value of n is, instead of (28) on page 832, approximately

$$k^2 = 3(n^2 + 1). \tag{28'}$$

Aperiodicity of the primary oscillation can be attained with this stipulation if $n^2 > 8$.

higher as k is larger. We will obtain the most favorable effect if we determine k so that the point n^2 lies directly on the transition arc between the two regions of the relevant curve. For the special case of our Fig. 131 ($v^2 = 1$), this transition arc contracts into the branch point with the parameter value $n^2 = k^2/4$. *We therefore obtain*, in reverse, *for fixed* n^2 *and for* $v^2 = 1$,

$$k^2 = 4n^2 \tag{28}$$

as the most favorable braking strength.

8. Numerical example for the "Silvana." In order to obtain an indication of the practically relevant magnitudes for the values of the quantities v^2, k^2, n^2, we consider the example of the stabilizer for the "Silvana," which was mentioned at the beginning of the previous section.

The properties of the "Silvana" are

Water displacement (Q):	850 000 kg
Metacentric height (H):	0,4 m
Free period of the rolling motion $\left(\frac{2\pi}{\alpha_0}\right)$:	8 sec.

These values give

Moment of inertia (J):	550 000 mkgsec^2.

The approximate dimensions of the stabilizer are

Weight of the rotor and shaft (q):	6 000 kg
Moment of inertia of the rotating masses (Θ) :	175 mkgsec^2
Angular velocity of the rotor (ω):	189 sec^{-1}
Estimated moment of inertia about the transverse axis (j):	150 mkgsec^2.

Our parameter n^2 is thus

$$n^2 = \frac{N^2}{Jj\alpha_0^2} = \frac{N^2}{jQH} = \frac{175^2 \cdot 189^2}{150 \cdot 85 \cdot 4 \cdot 10^3} = 21.$$

Because of the strong damping, the frequency ratio v could not be determined experimentally. We may assume, however, according to a communication by letter from Mr. Otto Schlick, that $v < 1$.

Detailed information about the strength of the braking is unknown; in any case, the free pendulum oscillation of the stabilizer is aperiodically damped by the bearing friction, as we likewise learn from Mr. Schlick. Thus it follows that $k \geq 2$ in the case $v \leq 1$, since, according to the fundamental equation (2), $\sqrt{v^2 - k^2/4}$ is proportional to the frequency of the pendulum oscillation of the stabilizer. With additional braking, k must be all the more greater than 2. Formula (28) or (28′) would give $k = 9$ or $k = 8$ as the most favorable value. In the following we will assume $k = 7$. The previous Fig. 130 is calculated according to the method of No. 4 for this value of k, and the incrementally found points with the corresponding values of n are registered. The value $k = 7$ corresponds to a braking strength $w = 700$ mkgsec.

Fig. 130 shows that aperiodicity of the primary oscillation already occurs, in this case, at relatively small values of n^2 for all values $v^2 \leq 1$ (curves I, II, III). (The inscribed numbers always denote the value of n^2 corresponding to the points.) For the value $v^2 = 1{,}5$ (curve IV), the primary oscillation attains the boundary between periodicity and aperiodicity at $n^2 = 9{,}3$, and for greater n^2 again becomes periodic. For still greater v^2 (V, VI), the damping achieves a maximum value and then decreases, without the oscillation becoming aperiodic.

§6. Results and practical experience with the ship stabilizer.

We now turn to a short discussion of the results contained in the preceding section, and, in particular, their comparison with the experience of Schlick himself in his experiments with the "Seebär" and the practical implementation of the stabilizer construction for the "Silvana" and the "Lochiel."

In addition to earlier publications (cf. p. 808), the essential results are compiled in a recent presentation*) by Schlick. Mr. Schlick also had the kindness to disclose more extensive experimental results to us by letter.

It may initially be doubted that the starting point in §4, formulated in many respects in the interest of analytical simplicity, corresponds to reality. We began with the assumption of small oscillations, and indeed for both the ship and the stabilizer. For what concerns the ship, its oscillations are in fact small, especially after the activation of the stabilizer; the deflections amount to at most a few degrees. Without the stabilizer, the deflections of the "Silvana" in stormy weather were 10–15°. It is otherwise with the deflections of the rotor frame, which indeed, according to the remarks on p. 803, is considerably larger than that of the ship, as experiments completely confirm. In one case that we will shortly discuss, rotations as large as 45° to each side occurred.

*) Jahrbuch der Schiffbautechnischen Gesellschaft, 1909.

Under these circumstances, the assumption of small oscillations must cause the effect of the stabilizer to appear *more favorable* than it would in a rigorous calculation. For we assumed in the calculation of the gyroscopic effect produced by the rolling motion, which causes the stabilizer to oscillate, that the axis of the rotor was perpendicular to the long axis of the ship. According to formula (II) in §1, however, this gyroscopic effect diminishes when the stabilizer oscillates, and vanishes completely when the axis has turned by 90°, since it is then directly parallel to the axis of the rolling motion. We have therefore overestimated the strength of the coupling between the ship and the stabilizer, and thus the calculated influence of the stabilizer on the rolling motion must generally be too large. Since rotations only up to 45° occur, and since this rotation is attained in practice only rarely, the overestimation of the gyroscopic effect will not be considerable. In fact, it is well known in many cases that results found under the assumption of small angles of rotation also agree completely with experiments for finite rotations.*) Moreover, a rigorous calculation would be out of place here because of other uncontrollable circumstances.

In the practical implementation, deflections of the rotor frame greater than 45° were prevented by a "stopper," which therefore temporarily annulled the gyroscopic effect.

A further analytic simplification lies in the assumption $-k\frac{d\vartheta}{dt}$ for the rotor braking, which implies the existence of a braking force only in the presence of an oscillatory motion. In fact, the very heavy rotor frame has, even with disregard of the hydraulic brake and use of ball bearings, considerable bearing friction that follows the law of sliding or rolling friction for small velocities (Chap. VII, §2). According to this law, a certain finite moment is necessary to transform the frame of the rotor from rest into motion. Thus the rotor frame is not restricted, when the ship is stationary, to a vertical equilibrium position with its center of gravity at the lowest point; it can, rather, persist in equilibrium for a displacement with respect to the vertical up to a certain angle ϑ_0, which is determined by the dimensions of the frame and the magnitude of the "coefficient of static bearing friction," and which can

*) As the general basis for this, one can cite the fact that the sine and the arc by which it is replaced coincide in the second decimal place up to 20°, and in the first decimal place up to 45°.

be called the angle of friction of the system. This angle is rather large for the stabilizer, as is shown by Schlick's experimental attempts to make the stabilizer oscillate artificially. It is to be noted, on the other hand, that the assumption $-k\frac{d\vartheta}{dt}$ for the hydraulic brake is only a general schema, and expresses only the essential phenomenon of a braking force that always opposes the instantaneous motion. The particular form of the law is rather unimportant for the effect, as indeed the same error is committed without severe consequences in many physical problems.

Another source of error that must be considered in the treatment of all technical problems, particularly those that concern such high velocities as are present in the Schlick stabilizer, lies in the elastic deformability of the material. The direction of its influence is clear in this case: a further weakening of the calculated effect. For, on the one hand, the gyroscopic effect itself depends on the inner reactions of the assumed rigid rotor, and, on the other hand, the effect is carried over to the assumed rigid ship only by means of the reaction forces of the stabilizer. Both types of reactions will be diminished by the elastic compliance of the stabilizer and the ship material, even if it is small, so that the effect of the stabilizer coupling must generally be reduced.

We will now seek to estimate, on the basis of the results of the previous section, the impulse strength required for the achievement of the most favorable effect, but can naturally expect, according to the preceding critical remarks, only an approximate indication.

We make the question more precise as follows.

How large must the impulse of the stabilizer be chosen so that the amplitude of the forced oscillation is reduced to an appropriate fraction of its original value (that is, the value without the stabilizer)? The impulse and the braking k should be sufficiently strong that the free oscillations, which are always re-excited by the initiation of the forced oscillations, are quickly damped.

The answer is to be taken from the deliberations of §5. If we restrict ourselves to the essentially present case of resonance between the waves and the free oscillation of the ship, to which the experiments of Schlick also refer (that is, $\gamma = 1$), then the desired relation will be given with sufficient approximation by equations (17) (p. 810) and (18) (p. 811) as

$$\left|\frac{A_n}{A_0}\right| = \frac{K\sqrt{k^2 + (1 - v^2)^2}}{n^2}.$$

We further note that according to the conclusions of p. 825, v^2 should be smaller than or at most equal to 1, a result that Schlick also obtained from his experiments; we can thus set, in that we neglect the proper fraction $(1 - v^2)^2$ in comparison with the large number k^2,

$$\left|\frac{A_n}{A_0}\right| = \frac{Kk}{n^2}. \tag{29}$$

An approximate value of K can be deduced from the experiments of Schlick with the "Seebär." It occurred there (see also Fig. 133) that the deflection of the ship in still water was reduced without the stabilizer to approximately $\frac{1}{20}$ of its original value after 20 oscillations. The corresponding value of K is

$$K = 0{,}05.$$

It is favorable to choose the quantity k, according to the results of the previous section, so that the free oscillation of the ship can be damped aperiodically by the coupling of the stabilizer, and indeed for relatively small values of the impulse. This condition is approximately satisfied for the value $k = 7$ that is adopted as the basis of Fig. 130. With consideration of the previously mentioned indeterminacy of the definition of k, $k = 7$ to $k = 10$ is chosen, corresponding, for the "Silvana" (p. 832), to a braking strength $w = kj\sqrt{QH/J}$ of 700 to 1000 mkgsec.

According to p. 830, a somewhat smaller value would indeed suffice to make the primary oscillation aperiodic, but the secondary oscillation would then, at least for $v = 1$, approach resonance with the primary oscillation; this can be avoided by increasing the strength of the braking. In order to achieve aperiodicity of the primary oscillation for $k = 7$, we must choose $n^2 = 10$ in the case $v^2 = 1$, and for $k = 10$, in contrast, $n^2 = 16$. For smaller values of v^2, strong damping of the oscillation of the ship will likewise be achieved by this choice (cf. Figs. 130 and 131). Through exclusive consideration of the free oscillation and its damping properties, we are therefore led to the values

$$n^2 = 10 \text{ to } 16, \text{ for } k = 7 \text{ to } 10.$$

We must now ask, further, whether the forced oscillations are also sufficiently influenced. We obtain from (29), by substitution of the established values of K, k, and n^2,

$$\left|\frac{A_n}{A_0}\right| = \frac{0{,}05 \cdot 7}{10} = 0{,}035$$

$$\text{or} \quad \left|\frac{A_n}{A_0}\right| = \frac{0{,}05 \cdot 10}{16} = 0{,}031.$$

Thus the values

$$n^2 = 10 \text{ to } 16$$

would suffice to reduce the amplitude of the ship oscillation to approximately 1/25 of the value that it assumes without the stabilizer. If Θ signifies the moment of inertia of the rotating mass about its axis, then the required angular velocity of the rotor is given, according to p. 804, by

$$\Theta\omega = 3{,}2\sqrt{QHj} \text{ to } 4\sqrt{QHj}. \tag{30}$$

For the above example of the "Silvana," there follows an angular velocity of 140 to 175 $\sec^{-1}$.*)

How do the experimental results of Schlick compare with this calculation? On the basis of the introductory remarks of this section, it is to be expected that the results will be slightly less favorable than those given by the calculation. Nevertheless, Schlick's experiments were astonishingly successful. Schlick found, for the "Silvana," a damping of the forced oscillation to $\frac{1}{10}$ of its original value with an angular velocity of the rotor of 189 $\sec^{-1}$. Since no details about the strength of the applied braking are known, this result cannot be immediately compared with ours. Schlick probably applied stronger braking than is assumed in our calculation, and it is already on this basis, in agreement with our results of p. 816, that we found a more favorable influence on the forced oscillation.

It is to be further noted that we began from the assumption of a harmonic wave that is directly in resonance with the oscillation of the ship, while in reality it is a matter of a superposition of a series of waves. Of this series, however, those not in resonance, according to the

---

*) Formula (30) gives, in our example, similar impulse strengths as a rough approximation derived by F ö p p l, which is intended to determine the order of magnitude of the required impulse if only the dimensions of the ship and its initial (to be annihilated) deflection (ψ_0) are known, assuming in addition that the frame of the rotor can oscillate up to 45°. The relevant formula is (cf. *Technische Mechanik*, Bd. VI, §41)

$$\Theta\omega = \frac{\psi_0}{5}\sqrt{QHJ}.$$

If the maximum value of ψ_0 is taken as approximately 15°, then

$$n = \frac{\Theta\omega}{\sqrt{QHj}} = \frac{1}{20}\sqrt{\frac{J}{j}};$$

if one therefore estimates the ratio J/j as for the "Silvana" (cf. p. 832), then one has approximately $n = \sqrt{10}$, as our above deliberation also gave.

developments of the first part of §5, are of less influence than that in resonance; for a certain range of frequency of the incoming waves, the stabilizer indeed strengthens the forced oscillation.

With consideration of all these circumstances, the approximate coincidence of our result with Schlick's provides a certain confirmation of the differential equations that were formulated as a starting point in §4.

We have not yet considered, however, an important point in the preceding calculation; namely, whether the amplitude of the rotor frame for the forced oscillations remains within the required bounds under the assumed conditions. Only then can an approximate agreement with practice be expected. The amplitude of the rotor frame is calculated from the formulas on p. 807. It follows from equations (15) and (17), again under the assumptions $v^2 = \gamma^2 = 1$ and small K, that

$$\left|\frac{a}{A_0}\right| = \left|\frac{a}{A}\right| \cdot \left|\frac{A}{A_0}\right| = \frac{K}{n}\sqrt{\frac{J}{j}},$$

and therefore, on the basis of our previously assumed numerical values and the impulse calculated above, approximately

$$\left|\frac{a}{A_0}\right| = 1.$$

The experimental amplitude A_0 of the ship without the stabilizer (cf. the diagrams in Fig. 132) is at most 10° to 15°. If one adds a correction to account for the no longer strictly applicable assumption of small oscillations, then the amplitude a will be approximately 20°, and therefore within the bounds for which our calculation is to be regarded as valid.

Schlick designed an indicator*) that produced an automatic graphical record of the rolling motion of the ship in his experiments. The indicator depended, in its turn, on the gyroscopic principle that the axis of a rotor supported at its center of gravity retains a fixed position in space. Experiments were conducted on the "Seebär," and later on the "Silvana" and "Lochiel" in stormy weather. The boat was first positioned abreast of the swells with the stabilizer held fast by the brake. The boat rolled up to 15° to each side, and, in individual cases for the "Seebär,"

*) Mentioned in the presentation to the Schiffbautechnische Gesellschaft (p. 135).

even up to 25°. The stabilizer was then released, and the rolling motion was almost instantaneously reduced to 1° to 2°. The moment of the stabilizer release corresponds in the diagrams to point A (see Fig. 132), after which follow only the small rolling motions that are shown in the diagrams.*)

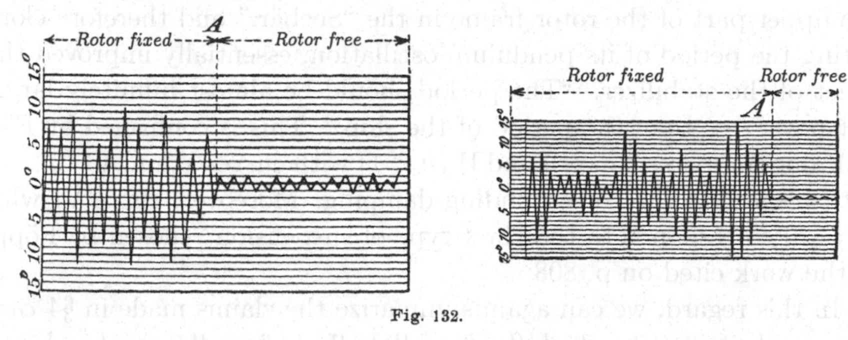

Fig. 132.

A further series of experiments was conducted, especially on the "Seebär," in order to measure the damping of the free oscillations under the influence of the stabilizer. In these experiments, a crane was used to give the boat an artificial lateral inclination of about 15°, and the number of oscillations until the inclination was reduced to $\frac{1}{2}°$ was observed. The diagram (Fig. 133) shows the effect of the stabilizer. Curves 1 and 2 correspond to the decrease of the inclination for the fixed stabilizer, and curves 3, 4, 5, 6 for the free stabilizer. In curve 5, for example, the inclination decreases from $15\frac{1}{2}°$ to $\frac{1}{2}°$ in four oscillations, while it decreases without the stabilizer in curve 2 from $13\frac{1}{2}°$ to $\frac{1}{2}°$ in 25 oscillations. We indeed concluded from the analytical treatment that it is possible to achieve aperiodicity in the oscillation of the ship for a relatively small impulse, and in particular for $v^2 \leq 1$, when the pendulum oscillation of the stabilizer is not faster than the oscillation of the ship. Whether the damping can be

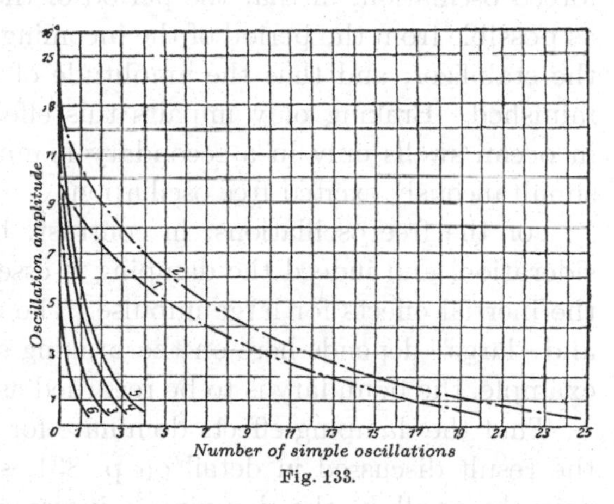

Fig. 133.

*) The figures are reproduced from the cited presentation. Similar diagrams are found in the article in the Ing.-Zeitschr. 1906, p. 1933.

elevated to this degree in practice remains undecided; the curves of Fig. 133, in any case, are not far from aperiodicity.*)

That $v^2 \leq 1$ is the most favorable arrangement coincides thoroughly with experience, and was also deduced by Schlick himself from his experimental results. We take from a communication by letter that loading the upper part of the rotor frame in the "Seebär," and therefore elongating the period of its pendulum oscillation, essentially improved the effect of the stabilizer. "The period should be almost infinitely large, but never less than the period of the ship." This is confirmed by Fig. 130: one compares curves I and II ($v < 1$) with curves V and VI ($v > 1$) with respect to the corresponding damping. Moreover, curve VI with $v^2 = 20$ corresponds to the "first type of suspension" chosen by Föppl in the work cited on p. 808.

In this regard, we can again summarize the claims made in §4 *concerning the extent to which the "inertial effects," on the one hand, and the "damping effects," on the other hand, come into consideration for the design of the stabilizer.* According to the previous discussion, the "resonant effects," and thus the inertial effects, are essential for the forced oscillation, in that the period of the ship is distanced as much as possible from the period of the incoming waves by the coupling with the stabilizer, and thus the amplitude of the forced oscillation is diminished. Braking only impairs this effect. Braking acts favorably in ocean swells only in a secondary manner, by damping the always simultaneously excited free oscillation.

For the free oscillations, in contrast, both effects come into consideration, and indeed the damping is essential for small impulse and the inertial effects for large impulse. The distinction between "small" and "large" depends here on the braking strength k. (For $v^2 = 1$, for example, the boundary is to be regarded as approximately $n^2 = k^2/4$.)

That the damping effects dominate for small n^2 first emerges from the result discussed in detail on p. 831, since, for small n^2 and sufficiently small k, the damping s increases with k, so that the primary oscillation can even become aperiodic, while for $k = 0$ the primary oscillation always remains undamped. On the other hand, the "free frequency" $\sqrt{p}$ of the primary oscillation becomes larger here,

*) One could expect, according to a remark of Föppl, an improvement of the braking effect by a hand regulation of the brake during the operation, in order to take into account that the top must possess, on the one hand, sufficient freedom of motion to take up the energy of the rolling motion, and must be braked with sufficient strength, on the other hand, to dissipate it.

at least in the favorable case $v^2 < 1$, and therefore the inertial effect is indeed undesirable in a certain sense. For large values of n^2, in contrast, we have seen that the damping effect can no longer increase with increasing n^2, since it approaches the damping of the ship as an upper limit. The inertial effects, on the other hand, increase continuously and become dominant.

In conclusion, we would not pass over certain doubts that have been raised against the construction of the stabilizer, particularly from the technical side.

They concern the question of whether the favorable stabilizing effect is not bound with other undesirable consequences on the ship. It was first conjectured by the marine engineer F. Berger*) that the stabilizer could be dangerous, since it "transforms rolling motion of the ship into pitching motion, and, in reverse, the latter into the former." When the stabilizer is strongly braked, as is indeed required for the achievement of a favorable effect, the energy of the pendulum motion is not completely dissipated in the brake, but rather a moment about the transverse axis is transmitted to the ship as a reaction to the braking forces and the gyroscopic effects, and therefore excites a pitching motion. If the frame of the rotor were fixed, for example, the gyroscopic effect produced by the rolling motion would act immediately on the ship instead of the frame of the rotor. A pitching motion of the ship, on the other hand, if it is transferred to the frame of the rotor by the brake, causes a rolling motion of the ship by the gyroscopic effect in the same manner as the pendulum motion of the frame. These two effects are undoubtedly present; that they are not damaging to the ship is due to their smallness.**)

1. Pitching excited by rolling. If one imagines, for example, that the stabilizer is completely fixed, in which case, as is easy to show analytically, the greatest moment will be carried over to the ship, then the two degrees of freedom of rolling and pitching, which are coupled by the gyroscopic effect, are qualitatively equally important; the amplitude of the latter, however, is much smaller than that of the rolling motion about the long axis because of the much larger moment of inertia about the transverse axis, and thus this "first Berger's effect" is certainly to be entirely neglected.

We can also take this result directly from the previous deliberations in §4. While we have thus far regarded the long axis of the ship as

*) Ztschr. d. V. d. Ing., 1906, p. 982.[305]

**) Cf. a remark of Föppl, ibid. 1906, p. 983.

fixed, its rotation about the transverse axis is now to be considered an additional degree of freedom. When the frame of the rotor is fixed, this degree of freedom directly takes the place of the pendulum motion of the frame, so that the coupling between the rolling and pitching motions is described qualitatively by our previous equations. While we previously emphasized that the fixed rotor had no influence on the rolling motion, this is now no longer the case, in that the possibility of the pendulum motion is replaced to a small degree by the pitching. For the moving rotor frame, on the other hand, the oscillation is naturally distributed between the pendulum motion and the pitching motion in a ratio that is determined by the strength of the braking.

For a fixed rotor frame, however, the moment of inertia for the coordinate ϑ, which now measures the pitching oscillation instead of the pendulum motion, is far greater than the moment of inertia of the rolling motion. In the place of the previous result (cf. p. 803) that an initial pendulum motion of the stabilizer can excite only a very small rolling motion except in the case of resonance, the assertion now is that a rolling motion can cause only a small pitching motion, unless the rolling and the pitching motions stand in direct resonance.

2 Rolling excited by pitching. The second effect, in contrast, may deserve a more detailed quantitative examination, again on the basis of the formulas of §4, since its significance has often been discussed. We imagine, for example, that the ship is initially at rest with an inclination angle ϑ_0 about the pitching axis, and ask for the following motion, and in particular for the fraction of the energy that is intermittently transformed into energy of rolling motion. The course of the motion is described, since we again imagine that the rotor frame is fixed to the ship, by the formulas (8‴) of §4 (p. 802), if only we now replace the pendulum motion of the stabilizer assumed there by the pitching motion of the ship; that is, replace the moment of inertia j by the moment of inertia K of the ship about its transverse axis, and the gravitational moment $q \cdot h$ by $Q \cdot S$, where S signifies the metacentric height for the pitching oscillation. The motion in each degree of freedom is naturally again the superposition of two oscillations, and indeed the oscillations will now differ much less in their frequencies, for a given impulse of the stabilizer, than the pendulum motion of the stabilizer did previously from the free rolling and pitching oscillations without the stabilizer.

Under the exclusion of absolute resonance between the rolling and pitching motions, we can thus extract an approximate representation of the resulting pitching motion from formulas (8‴). We replace the frequencies α_1 and α_2 by their values in the case $N = 0$, and obtain

$$\vartheta = a_1 \cos\sqrt{\frac{QH}{J}}\,t + a_2 \cos\sqrt{\frac{QS}{K}}\,t,$$

$$a_1 + a_2 = \vartheta_0,$$

$$\left|\frac{a_1}{a_2}\right| = N^2 \frac{SJ}{Q(KH - JS)^2}.$$

If we further introduce, exactly corresponding to our previous parameters n^2 and v^2, the dimensionless quantities

$$\mathfrak{N}^2 = \frac{N^2}{QHK}; \quad w^2 = \frac{QS/K}{QH/J},$$

then

$$\left|\frac{a_1}{a_2}\right| = \mathfrak{N}^2 \frac{w^2}{(1-w^2)^2}.$$

According to its definition, our current parameter $\mathfrak{N}^2$ is very small in comparison with the previous parameter n^2; in particular (cf. p. 832),

$$\mathfrak{N}^2 = n^2 \cdot \frac{j}{K}.$$

Now K, the moment of inertia about the pitching axis, will be at least 50 times the moment of inertia J about the rolling axis, and for the "Silvana," according to p. 832, $J = 4000j$, approximately; therefore

$$\mathfrak{N}^2 = 2n^2 \cdot 10^{-5}.$$

On the basis of the parameter value $n^2 = 20$ chosen for the "Silvana,"

$$\mathfrak{N}^2 = 4 \cdot 10^{-4}.$$

On the other hand, the frequency factor $w^2/(1 - w^2)^2$ does not exceed, with exclusion of resonance, the order of magnitude of $|a_1/a_2|$, in that it is greater than 1 only in the next approximation from $w = 1$.

From this ratio, however, we can extract in a well-known manner the fraction of the energy that is intermittently drawn from the pitching motion and transformed into the rolling motion. The superposition of the two oscillations is, namely, a "beat" in the pitching motion, with the maximum energy*)

$$E_1 = \frac{QS}{2}(a_1 + a_2)^2 = \frac{QS}{2}\vartheta_0^2,$$

and the minimum energy

$$E_2 = \frac{QS}{2}(a_1 - a_2)^2.$$

*) These formulas are admittedly derived under the assumption that the frequencies α_1 and α_2 approximately coincide, and thus w is little different from 1. In a more precise calculation, a power of w would enter as a factor in the expression for the oscillation of the energy; this does not, however, affect the order of magnitude.

The oscillation of the energy therefore amounts to only 10^{-4} to 10^{-3} of the mean energy of the pitching, and this fraction will only intermittently be transformed into rolling energy. We thus arrive at the result that the "second Berger's effect," excluding the case of resonance between the rolling and the pitching motions, will not seriously come into question for the ship.

Moreover, Mr. Skutsch*) has suggested making the discussed effect harmless by the use of a second stabilizer. The two stabilizers should be installed symmetrically with respect to the central transverse axis with opposite sense of rotation but equal rotational speed, and should be exact mirror images in all respects, which is to be achieved by constrained coupling. Because of the opposite rotational senses, opposite deflections of the two stabilizers would be produced by the rolling motion, and the pitching moments transmitted to the ship would thus cancel. The same holds, in reverse, for the moments produced by the pitching motion.

3. Material requirements. A further doubt that has been raised against the application of the stabilizer rests on the assumption that a ship which is hindered from rolling in the depicted manner must suffer especially strongly at sea, particularly from overbreaking waves. It is stated by the participants of the experimental voyages, however, that only a gentle rising and falling of the ship on the wavy surface occurred in place of the rolling motion, without overbreaking waves; the rolling waves simply vanished under the ship. On the other hand, it would well be imaginable that the hindering of the rolling motion places stronger loads on the interior bracing. Schlick himself denied such doubts by pointing to the great loads that large ships have borne in ocean swells, which they indeed follow only little, even if the incoming waves stand in direct resonance with the oscillation of the ship. In the same manner, it is known that the rolling motion of a sailing ship under full sail is completely suppressed by air resistance without damage to the ship.

One can further mention that the entire energy of the rolling motion to be dissipated by the stabilizer is by no means particularly large. Otherwise it would not be generally possible to dissipate the entire energy by a stabilizer that is relatively small in proportion to the ship.

*) Ztschr. d. V. d. Ing. 1908, p. 464.

We calculate, on the basis of the numerical values given for the "Silvana" on p. 832, the moment

$$N \cdot \frac{d\vartheta}{dt}$$

applied by the stabilizer to the ship, and, in particular, the moment corresponding to the oscillations that are forced by the ocean swells. We choose their period, corresponding to the mean conditions, to be $\tau = 8$ sec. If we assume the most unfavorable case in which the frame of the stabilizer achieves the greatest possible deflection of 45°, then $\vartheta = \frac{\pi}{4} \sin 2\pi \frac{t}{\tau}$; the maximum value of $\frac{d\vartheta}{dt}$ will thus be $\frac{2\pi^2}{4\tau} = \frac{\pi^2}{16}$ $\sec^{-1}$. According to p. 832, $N = \Theta\omega = 175 \cdot 189$ mkgsec. The moment in question will therefore be approximately 20 000 mkg. If we take the distance between the bearings of the rotor frame as approximately 2 m, then a force of 10 tons will act on each bearing, which is moderate in proportion to the otherwise present weights in the ship. This is the load that the ship experiences from the stabilizer.

The cited doubts are hardly an essential impediment to the extensive application of the ship stabilizer; the true problems in the further perfection of the construction generally lie not on the dynamic, but rather on the technical side.

The future must tell whether the bold idea of Schlick can be generally adopted in marine transportation in spite of the considerable costs and difficulties that oppose its development, or whether it will stand in history more as a striking example of the high standard for the actual implementation of a correctly conceived mechanical construction.[306]

§7. The gyroscopic compass.

In the discussion of Foucault's experiments to demonstrate the rotation of the Earth by its effect on the motion of a rapidly rotating top, we mentioned Foucault's idea*) that this effect could be used for orientation on the Earth, and thus for the development of a gyroscopic instrument as a compass. We considered the properties of a top whose figure axis may move only in the horizontal plane (a top "with two degrees of freedom"), which automatically aligns itself in the meridian like a declination needle, and a top whose figure axis may move only in the meridional plane, which aligns itself in the direction of the Earth's axis like an inclination needle.**)

*) p. 734 ff. **) p. 746 ff.

The practical implementation of Foucault's idea could naturally proceed only after it became possible to produce rotors that could spin for a long duration without interruption, which is now accomplished by the use of electric motors. An essential technical requirement here is that the external moment that opposes friction and maintains a constant eigenimpulse act strictly about the figure axis, and thus produce no impulse change in the equatorial plane. Experiments were conducted in the 1880s by the navies of France, Holland, and England, but few of their results were published; in any case, these experiments did not lead to the lasting introduction of a rotational compass.*) The first experiments in Germany were made in the 1890s by Werner von Siemens, in support of a patent filed by Van den Bos.**)[309]

The problem of constructing a practical gyroscopic compass has now become acute, particularly for use on warships, on which the magnetic field of the Earth is distorted by the increase in the mass of iron to such an extent that the reading of the magnetic compass becomes uncertain in the highest degree, and indeed almost unusable. The situation is especially difficult because the mass of iron on board (munitions, projectiles) is often displaced, and because the operation of electrical machines directly influences the compass needle magnetically; in addition, the induced magnetism on the ship from the Earth's field depends on the motion of the ship.***) Various firms have made experiments in this direction, of which, however, only those of Anschütz-Kämpfe in Kiel have so far attained to publication†) and led to practical application.

The Anschütz compass uses not the Foucault arrangement, but rather a top with three degrees of freedom under the influence of gravity, whose center of gravity lies either on the figure axis or in the equatorial plane of the top.

*) M. Edm. Dubois: Sur le gyroscope marin, C. R. 98, p. 227, 1884[307]; W. Thomson: Gyrostatic model of a magnetic compass, Nature 30, p. 524, 1884.[308]

**) D. R. P. Kl. 42, No. 34513.

***) Cf. Encyklopädie d. math. Wissensch. VI 1, 5, p. 320 ff. (Nautik, von H. Meldau).[310]

†) Jahrbuch der Schiffbautechnischen Gesellschaft, X, 1909, p. 352; math. appendix (Max Schuler), p. 561. The experiments of the firm Hartmann & Braun in Frankfurt a. M. after the design of Prof. Ach have not yet been published.[311]

If we completely disregard its weight, the top would be in equilibrium with respect to the Earth and with respect to space if its figure axis were parallel to the axis of the Earth. This equilibrium is perturbed by the effect of gravity; in the first case we have a heavy symmetric top, and in the second case, at least according to the applied forces, a heavy asymmetric top.

For the heavy symmetric top, gravity initially draws the figure axis from its spatially fixed position, and the resulting spatial motion causes further deflecting gyroscopic effects. An equilibrium position with respect to the Earth now becomes possible, as for the previously treated barogyroscope of G i l b e r t.*) The spatial motion of the top that is stationary with respect to the Earth is then, considered in space, a regular precession of the figure axis about the axis of the Earth in which the gravitational force and the gyroscopic effect maintain equilibrium. The figure axis lies constantly in the meridional plane of the observation location, at a determined angle with respect to the vertical. The determination of the angle is provided simply by the condition

gyroscopic effect of the Earth's rotation = gravitational moment

for the regular precession, with the gyroscopic effect given by formula (II) of p. 764.

In this precession, the support point of the top is not fixed, but rather is led, even if it be at rest with respect to the Earth, on a parallel circle about the axis of the Earth; it therefore experiences a constant radially directed acceleration in the plane of the parallel circle. An acceleration of the support point, however, acts as an oppositely directed force at the center of gravity (cf., for example, p. 612); the magnitude of this force is the product of the acceleration and the mass of the system. This apparent force is the centrifugal force corresponding to the rotation of the Earth, which also causes the oblateness of the Earth. We account for it correctly if we consider the actually measured force in the direction of the plumb line (the resultant of the pure gravitational force and the centrifugal force) as the gravitational force.

This ideal form of motion of the top, which we wish to designate from now on as "spatial precession," would occur if the axis of the top were initially placed in the meridian at exactly the correct elevation with respect to the horizontal; the placement would depend, however, on the geographical latitude and the velocity of the ship, and must therefore be corrected for its motion. On this principle rests a model of

*) p. 753.

a gyroscopic compass designed by A n s c h ü t z - K ä m p f e, which his firm declared in a patent at an earlier stage of their investigations.*)

If the top is to indicate the north–south line in its spatial precession, it must have a perceptible deviation from the vertical; thus the gravitational moment must be very small, because of the relative smallness of the gyroscopic effect caused by the rotation of the Earth. As a consequence, however, the equilibrium position in the meridian has a very small degree of stability. The top leaves its equilibrium position at each of the small impacts that continuously occur due to the motion of the ship, and executes, analogously to the case of the stationary Earth, a slow precession relative to the Earth about the vertical, or at least returns to the meridian in a very slow oscillation.

This insufficient stability is the primary reason that the first form of the compass has not advanced to implementation.

We now consider the case of the asymmetric top, which is used in the gyroscopic compass implemented by A n s c h ü t z - K ä m p f e.**) Let the outer ring of a Cardanic suspension rotate about the vertical and the inner ring rotate about the horizontal diameter of the outer ring, the "line of nodes" of the top. This inner ring carries an added weight in such a manner that the plane of the inner ring is *horizontal* in its equilibrium position; the *figure axis* whose bearings are in the plane of the inner ring perpendicular to its rotation axis is also *horizontal* in this equilibrium position. The center of gravity of the system then lies in the equatorial plane of the top, without, however, taking part in the rotation about the figure axis. This purely schematic image of the arrangement suffices for our present purpose;[313] the actual implementation will be discussed at the end of this section.

The system can still be treated kinetically as a symmetric top if the eigenrotation is so strong that the inertia of the rings may be neglected with respect to that of the rotor. The gravitational moment then acts, as before, about the line of nodes of the top, only it is no longer proportional to the sine, but rather to the cosine of the previously defined angle ϑ of the figure axis with respect to the vertical. The prior results

*) D. R. P. Nr. 178814, 1905.

**) D. R. P. Nr. 182855, 1908. Experiments with this system have also been made by O. M a r t i e n s s e n; cf. Physik. Ztschr. VII, Nr. 15, p. 535, 1906.[312] In the experiment of F ö p p l for the demonstration of the rotation of the Earth, the elasticity of the suspension wires for the likewise horizontally placed top takes the place of the gravitational force. Sitzungsber. d. Akad. d. Wiss. München, 1904, Heft I. Cf. also the addenda at the end of Vol. III.

for the symmetric top may thus be applied to this arrangement. A stationary "spatial precession" is possible in which the figure axis constantly points in the northern direction of the observation location, with a constant elevation above the horizontal. The elevation is again determined by the precessional condition

gravitational moment = gyroscopic effect,

and can be made very small if the gravitational moment is sufficiently large.

This stationary motion is transformed, in the case of the nonrotating Earth, into an equilibrium position with a horizontal axis. Similar to this limiting case of vanishing rotation of the Earth, the general motion is the superposition of the stationary motion with a precession about the vertical, a "horizontal precession" that occurs if a sufficiently large impact disturbs the spatial precession, but is naturally influenced in its character by the rotation of the Earth. For a small impact, in contrast, the horizontal precession will degenerate into an oscillation about the northern direction.

The origin of the directional force that pulls the top back to the meridian is not immediately evident here; we recognized such a force previously only for the Foucault top with two degrees of freedom. The dynamic operation of the gyroscopic compass can now be made clear by comparison with the Foucault top that moves only in the horizontal plane. The later analytic supplement will support this general representation in more detail.

In the case of Foucault, the guidance in the fixed horizontal plane prevents the axis from remaining stationary in space as the Earth rotates. The resulting forced spatial motion has the consequence of a gyroscopic effect that draws the axis back to the meridian, and thus causes an oscillation about the meridian. The displacement of the impulse in this oscillation is accomplished by the moment of the force that is perpendicular to the horizontal plane (cf. p. 746). What the forced guidance in the horizontal plane accomplishes for Foucault is effected in our arrangement by the gravitational moment of the added weight, which represents, to a certain degree, a compliant guidance. It accomplishes, in a qualitative sense, the same impulse displacement as the guiding force in the case of Foucault, and therefore draws the figure axis back into the equilibrium position. Since, moreover, the elasticity of the horizontal guidance also comes into consideration for the Foucault arrangement, the guidance there is actually not completely enforced, so that there is no qualitative difference between the two cases; the gravitational moment for Foucault is only to be imagined as very large.

If, for example, the "north pole" of the figure axis is initially horizontal and points to the east, it will at first retain its position in space, and thus be elevated with respect to the horizontal because of the rotation of the Earth. The resulting gravitational moment displaces the impulse from its eastern position to the north, and thus the figure axis, which follows the impulse axis in the mean, as is known from pseudoregular precession, experiences a directional force to the north. If, in contrast, the "south pole" of the figure axis initially points to the east, then this pole will, in the same manner, be deflected to the south; the north pole again experiences a directional force to the north, so that the north-pointing position is, in fact, a stable equilibrium position for the north pole of the figure axis.

The equilibrium position for the present arrangement indeed has, as the analytic developments will confirm, a relatively large degree of stability compared with the previously mentioned symmetric form, because of the increased gravitational moment that effects the positioning in the meridian in the depicted manner in a correspondingly shorter time. Compared with the Foucault arrangement, the increase of stability is of a different kind; it consists, namely, of an unresponsiveness to disturbances of not too long a period that is due to the presence of three degrees of freedom. The schema of this stabilization differs from that treated in §1 No. IV only by the addition of the gravitational moment, which restricts, in a certain measure, the degrees of freedom. The specific capability of resistance against changes of direction emphasized in §1 is expressed here as a slow yielding, an "effective inertia" that makes the top unresponsive to disturbances of not too long a period. The effective inertia vanishes only for an infinitely large gravitational moment, the case of the Foucault top, in which the disturbances are opposed only by the moment of inertia of the top itself. The stability can be further strengthened by appropriate damping devices, whose remarkable development has had a primary share in the success of the gyroscopic compass.

The greatest possible increase in the stability of the stationary north–south position is particularly required because of the influence of the ship's motion on the top. Every motion of the ship represents a disturbance of the equilibrium condition of the top, and thus causes oscillations that must be held within narrow bounds in order for the device to be practical. Two questions must now be answered: 1. How does the top react to a uniform motion of the ship? 2. How does the top react to an acceleration of the ship?

1. If the ship travels with constant velocity from east to west or vice versa, this motion is equivalent to a small decrease or increase in the rotational velocity of the Earth. In the first case, the precessional velocity of the stationary spatial precession, assuming that it has been established, is a little smaller than when the ship is stationary; this corresponds to a smaller gyroscopic effect, which therefore requires a smaller gravitational moment to maintain equilibrium. The consequence is only a somewhat smaller elevation of the figure axis above the horizontal. In the second case, in contrast, the figure axis will assume a somewhat greater elevation. This effect is the same as for a passage to another geographic latitude and is harmless, since it does not influence the pointing.

A motion in the north-south direction, on the other hand, is equivalent to an additional rotation of the Earth about an equatorial diameter; the resultant axis of the Earth's rotation and this added rotation deviates little, however, from the polar axis because of the smallness of the latter. The final effect is that the figure axis is placed in the vertical plane through this resultant rotation axis in the stationary spatial motion. Its direction therefore deviates from the meridional plane, but only, except at high latitudes, by a small angle. The deviation is to the west for south–north motion and to the east for north–south motion; it amounts for a mean velocity of 16 knots per hour to*)

latitude:	0°	20°	40°	60°	70°
deviation:	1,02°	1,04°	1,33°	2,03°	2,97°.

2. We now turn to the acceleration effects, which occur for the passage from one velocity to another that differs in magnitude or direction. Due to the damping that is intentionally included in the actual construction, the figure axis is positioned in the new equilibrium position as soon as the velocity again becomes uniform, if this position possesses a sufficient degree of stability. The disturbance of the original equilibrium condition, however, produces an oscillation whose amplitude can exceed the final deviation. How, in detail, do such oscillations come about?

For a change of velocity of the ship in the east–west direction, the top initially maintains its spatial precessional velocity, since the gravitational moment is initially unchanged. Since this velocity no longer

*) See p. 569 of the article by Schuler cited on p. 846.

coincides with the spatial velocity of the ship, the top deviates from the meridional plane laterally, and begins a slow horizontal precession with an overlaid nutation. Such a motion must always occur if the equilibrium condition of the spatial precession (p. 849) is disturbed, just as any impact produces a pseudoregular horizontal precession for a nonrotating Earth. A sufficient magnitude of the directional force toward the meridian is required in order that the figure axis does not execute a full precession, but rather returns with oscillations to the new equilibrium position in the meridian after a small deflection. We will have to make this requirement more precise, remarking in advance, however, that these deflections are very small, because of the smallness of the disturbance, in comparison with the (soon to be discussed) deflections caused by a velocity change in the direction of the meridian, and thus do not come into consideration in practice. Moreover, they are also less essential, since the course of a ship generally follows not a circle of constant latitude, but rather a great circle.

The acceleration of the support point is equivalent, according to the well-known law of inertia, to an opposite force at the center of gravity. For an east–west acceleration, its moment is directed about the figure axis and is thus inessential. For a north–south acceleration, in contrast, its moment acts about the line of nodes and has as a final consequence, like the gravitational moment, a pseudoregular horizontal precession that lasts as long as the acceleration continues. It is necessary that the total amount of this precessional motion remain within practical bounds for the duration of the acceleration. Moreover, this deviation is always directed toward the same side as the subsequently established new equilibrium position (cf. above), so that only the difference of these two deviations comes into consideration as the amplitude of the resulting oscillation.

Accelerations of the ship in the meridian and perpendicular to it come into question for any turn. Like the magnetic needle, the figure axis does not remain completely stationary in a curve, but rather makes small oscillations; the gyroscopic compass may correctly detect the curve only if these oscillations are sufficiently small.

In addition to intentional accelerations of the ship, however, all nonuniform motions such as oscillations of the ship and the support of the top also come into essential consideration. Since such disturbances always have short periods, they can be made harmless, according to the well-known resonance principle, if the oscillation period of the compass is large. This is achieved by the previously mentioned effective inertia of the Anschütz compass, without the necessity of large and technically undesirable masses. On the other hand, the magnitude of the oscillation period is naturally restricted by the requirement that the compass should not take too long to attain its equilibrium position after a large disturbance. For the Anschütz-Kämpfe device, an oscillation period of 70 minutes is chosen as a favorable mean; moreover, external disturbances are weakened as much possible in advance by mounting the entire system in a Cardanic suspension. The necessary length of the oscillation period is an apparent difficulty, since an oscillation with such a long period is not immediately recognized by the eye as an oscillation.*) We will see, however, that any oscillation is accompanied by a small rising and falling of the figure axis with an equal period. A sensitive spirit level on the compass rose can thus allow the helmsman to recognize the presence of an oscillation and the instantaneous mispointing of the compass. During a simultaneous turn, however, the level will be temporarily useless for this purpose, since its equilibrium position is then disturbed by the centrifugal force.

We must now represent the course of the motion analytically. For the calculation of the gyroscopic effect, we again make the assumption that the impulse axis coincides with figure axis,**) and therefore neglect the impulse that corresponds to the motion of the figure axis about the axis of the Earth and relative to the Earth with respect to the rotational impulse about the figure axis. As the applied moments, we must consider the gravitational moment and the gyroscopic effects that arise from the spatial motion of the figure axis.

Let ϑ denote the elevation of the figure axis above the horizontal, and positive ϑ signify an elevation of the "north pole," the side of the figure axis from which the rotation follows in the counterclockwise sense; let ψ represent the western deviation of the north pole from the northern direction SN. Let P be the gravitational moment for $\vartheta = \frac{\pi}{2}$; $P \sin\vartheta$ is then the gravitational moment for the inclination angle ϑ. Let B be the moment of inertia of the entire system about the vertical axis, A the moment of inertia about the horizontal line of nodes, and N the eigenimpulse. In addition to the mass of the rotor, the masses of the suspension are supposed to be included in A and B, in

*) Cf. the discussion after the presentation of Anschütz cited on p. 846.

**) As in §1, Nos. I and II of this chapter. Cf. also the previous approximation formulas for pseudoregular precession in Chap. V that depend essentially on this assumption.

so far as they take part in the motion of the axis of the rotor. A is constant, and B can be regarded as approximately constant for small elevations ϑ. The motor produces a constant value of N. Let ω be the angular velocity of the Earth, and φ the geographic latitude. We decompose the rotation of the Earth into the components $\omega \sin\varphi$ about the vertical and $\omega\cos\varphi$ about the horizontal north–south direction.

These rotation components give rise to gyroscopic effects whose axes stand perpendicular to the relevant rotation axes and the impulse axis. A simple geometric deliberation gives the values of these effects, calculated according to formula (II) of §1, as

$$K = N\omega(\sin\varphi\cos\vartheta - \cos\varphi\sin\vartheta\cos\psi)$$

about the line of nodes, and

$$K' = -N\omega\cos\varphi\cos\vartheta\sin\psi$$

about the vertical. The gyroscopic effects due to the motion relative to the Earth are

$$K_1 = N\frac{d\psi}{dt}\cos\vartheta$$

about the line of nodes, and

$$K_1' = -N\frac{d\vartheta}{dt}\cos\vartheta$$

about the vertical.[314] The equations of motion are thus

$$(1)\qquad \begin{cases} A\dfrac{d^2\vartheta}{dt^2} = N\omega(\sin\varphi\cos\vartheta - \cos\varphi\sin\vartheta\cos\psi) \\ \qquad\qquad + N\dfrac{d\psi}{dt}\cos\vartheta - P\sin\vartheta, \\ B\dfrac{d^2\psi}{dt^2} = -N\omega\cos\varphi\cos\vartheta\sin\psi - N\dfrac{d\vartheta}{dt}\cos\vartheta. \end{cases}$$

We now restrict ourselves to motions for which the elevation angle ϑ always remains very small, as corresponds to the intended operation of the gyroscopic compass, and therefore neglect ϑ with respect to finite quantities, while $N\omega$ is regarded as a small quantity of the first order. The justification for this assumption lies in the relative magnitude of the gravitational moment, which represents, as for the Foucault horizontal top, a compliant guidance in the vicinity of the horizontal plane. Equations (1) then become*)

$$(2a)\qquad A\frac{d^2\vartheta}{dt^2} = N\omega\sin\varphi + N\frac{d\psi}{dt} - P\vartheta,$$

$$(2b)\qquad B\frac{d^2\psi}{dt^2} = -N\omega\cos\varphi\sin\psi - N\frac{d\vartheta}{dt}.$$

*) In agreement with the fundamental equations (4) in the cited article by O. Martienssen, Physik. Ztschr. Bd. 7 (1906), p. 535.

We obtain from (2a), by differentiation and substitution of $\frac{d\vartheta}{dt}$ from (2b),

$$(3) \qquad NA\frac{d^3\vartheta}{dt^3} = (N^2 + PB)\frac{d^2\psi}{dt^2} + PN\omega\cos\varphi\sin\psi.$$

On the other hand, a twofold differentiation of (2b) gives

$$(3a) \qquad -NA\frac{d^3\vartheta}{dt^3} = AB\frac{d^4\psi}{dt^4} + NA\omega\cos\varphi\frac{d^2\sin\psi}{dt^2}.$$

We would obviously obtain, by addition of (3) and (3a), an oscillation equation of the fourth order for the coordinate ψ, which therefore signifies a superposition of oscillations, and indeed, as we immediately add here, a slow and a fast oscillation that we can conceive as a precession and a nutation. These two oscillations are, in turn, overlaid with the spatial precession, which must be expressed by constant values of ψ and ϑ, and indeed, according to equations (2), by the values

$$\psi = 0, \qquad \vartheta = \frac{N\omega\sin\varphi}{P}.$$

The figure axis therefore points in this equilibrium position to the north, under a small elevation with respect to the horizontal, since $N\omega$ is always small and P can be chosen as sufficiently large.

We now consider the two oscillations separately. For the slow oscillation about the equilibrium position, equation (3) is approximated in a well-known manner by the equation

$$(4) \qquad (N^2 + PB)\frac{d^2\psi}{dt^2} + PN\omega\cos\varphi\sin\psi = 0,$$

which, because of the relative smallness of $PN\omega$ and bigness of N^2, actually represents a slow oscillation.*)

This slow oscillation is overlaid with rapid nutations that are obtained by the consideration of the previously neglected term. If one sets, namely,

$$\psi = \psi_0 + \psi_1,$$

*) If we take, namely, the magnitude of the left-hand side of (3) from (3a), then it is expressed by two terms that are individually small with respect to the two terms on the right-hand side. For $AN\omega\frac{d^2\sin\psi}{dt^2}$ is to be directly neglected with respect to $N^2\frac{d^2\psi}{dt^2}$, and $AB\frac{d^4\psi}{dt^4}$ is of the order $\frac{A\omega}{N}PB\frac{d^2\psi}{dt^2}$ for the slow oscillation (4), and is therefore to be neglected with respect to $PB\frac{d^2\psi}{dt^2}$. Thus the solution of equation (4) also represents, in fact, an approximate solution of equation (3).

where ψ_0 should satisfy equation (4), then one obtains for small deviations ψ_1, by elimination of ϑ from (3) and (3a), the equation

$$(5) \qquad AB\frac{d^4\psi_1}{dt^4} + (AN\omega\cos\varphi\cos\psi_0 + N^2 + PB)\frac{d^2\psi_1}{dt^2} + PN\omega\cos\varphi\cos\psi_0\cdot\psi_1 = 0.$$

If one sets $\psi_1 = ae^{\lambda t}$, then λ must satisfy the equation of the fourth degree

$$(5a) \qquad AB\lambda^4 + (AN\omega\cos\varphi\cos\psi_0 + N^2 + PB)\lambda^2 + PN\omega\cos\varphi\cos\psi_0 = 0.$$

For sufficiently large values of N, this equation is always satisfied by a large negative value of λ^2, which one obtains with sufficient precision, by neglect of the constant term in (5a), as

$$\lambda_1^2 = -\frac{N^2}{AB}.$$

This large frequency corresponds, in fact, to the rapid nutations of the figure axis for pseudoregular precession. The other two roots of (5a), in contrast, are small, and correspond to the previously assumed slow oscillation.

The superposition of the two oscillations represents the complete approximate solution of the system (2).*) For the rapid nutations, the oscillation in ϑ remains of the order of magnitude of the oscillation in ψ_1, as is well known from pseudoregular precession. We must still confirm that the angle ϑ also remains small for the slow oscillation, because of the relatively large gravitational moment that presses the figure axis into the horizontal plane. From (2) there follows

$$-\frac{d\vartheta}{dt} = \omega\cos\varphi\sin\psi + \frac{B}{N}\frac{d^2\psi}{dt^2}.$$

Here ω is small in itself, and $\frac{B}{N}\frac{d^2\psi}{dt^2}$, at least for the slow oscillations (4), is also small; ϑ therefore remains within narrow bounds. The visible character of the oscillation will therefore be a slow horizontal oscillation that is accompanied by a vertical oscillation with an equal period and a very small amplitude. Because of the smallness of $d^2\psi/dt^2$, the two oscillations are displaced in phase, according to (2b), by a quarter wave-

*) We would naturally have obtained this superposition of two oscillations immediately if we had also assumed ψ as small in equations (2a) and (2b). This assumption, however, is not always justified (ψ may be large, for example, in the initial position of the gyroscopic compass).

length, so that the apex of the top describes a flat ellipse, with the major axis in the horizontal plane and the minor axis in the meridian. The latter is so small, however, that the figure axis in fact appears as a horizontally oscillating magnetic needle.

Since the rapid nutations, which in reality are not perceptible because of their rapid decay, are neglected for this oscillation process, we have here, corresponding to our previous claims, a horizontal precession that is made nonuniform by the rotation of the Earth. One can conceive it as a slowly varying regular precession in which, because of the changing elevation, the changing gravitational moment maintains equilibrium with the change of position of the impulse. It is overlaid with the spatial precession and thus becomes nonuniform, while for vanishing rotation of the Earth it is transformed into a uniform horizontal precession with constant elevation.

According to equation (4), which is written in the dimensions of an oscillation equation as

$$\left(\frac{N^2}{P} + B\right)\frac{d^2\psi}{dt^2} + N\omega\cos\varphi\sin\psi = 0, \tag{4}$$

the top oscillates like a magnetic needle with the "directional force" (actually the moment of the directional force) $N\omega\cos\varphi$. The inertia of the needle corresponds here to an effective inertia that is composed of the moment of inertia B and the "inertia of the horizontal precession" $\frac{N^2}{P}$. The latter vanishes for infinitely large P, when the figure axis is restricted to the horizontal plane, as it is for the Foucault top with two degrees of freedom; equation (4) is then transformed into the oscillation equation for that horizontal top (see p. 748, equation (4)). This effective moment of inertia, which for sufficiently large eigenimpulse far overwhelms the original moment of inertia B, makes the top with three degrees of freedom insensitive in an elevated measure to external disturbances of not too long a period.

The period of the motion is, for small oscillations,

$$T = 2\pi\cdot\sqrt{\frac{N^2 + PB}{PN\omega\cos\varphi}}; \tag{6}$$

it therefore increases for large values of the eigenimpulse as $\sqrt{N/P\omega}$. One thus has, in the choice of the eigenimpulse and the gravitational moment, a means to regulate the oscillation period, while the magnitude of B is inessential in practice.

In order to make the gyroscopic compass insensitive to external disturbances, a quite long oscillation period is required, which is achieved here by a large value of the impulse. In contrast, an elongation of the oscillation period for the Foucault gyroscope could be achieved only by an undesirable increase of the moment of inertia B or by impermissibly small values of the impulse N.

Equation (4), however, describes an actual oscillation only for sufficiently small impacts, as is well known for the pendulum. The previously posed question of the impact for which a sustained uniform horizontal precession occurs, and the gyroscope does not return with oscillations to the equilibrium position, is identical with the analogous question for the simple pendulum:

"Is the motion a periodic oscillation, or does the pendulum reach its highest point and thus move constantly in the same sense?"

The answer is provided, in both cases, by the energy theorem. After multiplication by $\frac{d\psi}{dt}$ and integration, there follows from (4)

$$\left(\frac{N^2}{P}+B\right)\left[\left(\frac{d\psi}{dt}\right)^2-\left(\frac{d\psi}{dt}\right)_0^2\right]=2N\omega\cos\varphi\,(\cos\psi-1), \tag{7}$$

where $\left(\frac{d\psi}{dt}\right)_0$ is the angular velocity for $\psi=0$. The figure axis does not make a complete revolution (that is, ψ does not reach the value π) if

$$\left(\frac{d\psi}{dt}\right)_0^2<\frac{4PN\omega\cos\varphi}{N^2+PB},$$

or

$$\left(\frac{d\psi}{dt}\right)_0<\frac{4\pi}{T}. \tag{8}$$

Practically, however, this requirement is much too broad, since it is well known that the oscillation period for large deflections, as for the pendulum, is no longer to be measured by (6), but rather is much longer, indeed becoming infinite for the limiting case $\psi=\pi$. Thus only much smaller deflections than those given by (8) are permissible.

In the case of an east–west change of velocity of the ship, $\left(\frac{d\psi}{dt}\right)_0$ can immediately be deduced kinematically. Let $\Delta\omega$ be the additional angular velocity about the axis of the Earth that corresponds to the change of the ship's motion, which we imagine, for simplicity, as instantaneous. According to equation (2a), there holds after the change of velocity, with the neglect of the nutation and thus the term $A\vartheta''$,

$$N(\omega+\Delta\omega)\sin\varphi+N\frac{d\psi}{dt}-P\vartheta=0.$$

Since equilibrium obtained before the change of velocity,

$$P\vartheta - N\omega\sin\varphi = 0,$$

and thus, since ϑ initially remains unchanged,

$$\left(\frac{d\psi}{dt}\right)_0 = -\Delta\omega\sin\varphi;$$

that is, the figure axis initially continues the spatial precession with its velocity before the acceleration, and correspondingly deviates from the meridian.

The (indeed too broad) condition (8) that the top return to the meridian thus becomes

$$\Delta\omega\sin\varphi < \frac{4\pi}{T}.$$

For the Anschütz instrument, $T = 70$ min. is chosen, while $\Delta\omega$ is a small fraction of the angular velocity of the Earth. If we assume, for example,

$$\Delta\omega = \frac{1}{40}\omega = \frac{1}{40}\frac{2\pi}{24\cdot 60}\ \text{min}^{-1}$$

and take the geographic latitude as 45°, then

$$\Delta\omega\sin\varphi = \frac{1}{80}\frac{2\pi}{24\cdot 60}\sqrt{2}\ \text{min}^{-1} = \frac{4\pi}{160000}\ \text{min}^{-1},$$

while

$$\frac{4\pi}{T} = \frac{4\pi}{70}\ \text{min}^{-1}.$$

Our condition (8) is therefore thoroughly fulfilled. There is thus no danger that a west-to-east or east-to-west acceleration may give rise to a horizontal precession with a substantial deflection.

This deflection ψ_b is actually to be obtained from the energy equation (7) and the calculated initial velocity $(d\psi/dt)_0 = \Delta\omega\sin\varphi$. (The notation ψ_b indicates that the acceleration is along a circle of constant latitude.) If we set $(d\psi/dt)_{\psi=\psi_b} = 0$, there follows, approximately,

$$\cos\psi_b = 1 - \frac{N}{2P\omega}\frac{\sin^2\varphi}{\cos\varphi}\Delta\omega^2,$$

or, since φ_b is a small angle,

$$\cos\psi_b = 1 - \frac{1}{2}\psi_b^2,$$

where

$$\psi_b = \Delta\omega\sin\varphi\sqrt{\frac{N}{P\omega\cos\varphi}}.$$

The change $\Delta\omega$ in the angular velocity corresponds to the linear velocity change

$$\Delta v = \Delta\omega R\cos\varphi,$$

where R is the radius of the Earth. Thus

$$\psi_b = \frac{\Delta v}{R} \operatorname{tg} \varphi \sqrt{\frac{N}{P\omega \cos \varphi}}.$$

We now calculate the ballistic deflection, which corresponds to the velocity change Δv in the direction of the meridian, likewise assumed to be instantaneous. Such a change first exerts, according to p. 852, a moment that acts about the line of nodes. Its final effect is a horizontal precession with retention of the vertical elevation, at least to a good approximation; the top returns to the equilibrium position with oscillations after the cessation of the acceleration, without a continued precession. It is therefore necessary only to retain the entire deflection within sufficiently narrow bounds.*)

If Q is the moment of the acceleration pressure about the line of nodes, calculated according to p. 847 as

$$Q = -\gamma \cdot \frac{P}{g},$$

where γ is the acceleration, then Q is added to the gravitational moment $P\vartheta$ in equation (2a), while the term $A \cdot d^2\vartheta/dt^2$ is to be struck for the horizontal precession. Thus

$$N\omega \sin \varphi - P\varphi + Q + N\frac{d\psi}{dt} = 0.$$

The first two terms cancel here because of the equilibrium condition of the spatial precession, which is fulfilled for the elevation under consideration if the acceleration does not last too long. There remains

$$\frac{d\psi}{dt} = -\frac{Q}{N};$$

thus the total deflection (ψ_m indicates the acceleration in the meridian) becomes

$$\psi_m = \int_{t_0}^{t_1} \frac{Q}{N}\, dt = \frac{P}{gN}\Delta v,$$

since

$$\Delta v = \int_{t_0}^{t_1} \gamma\, dt,$$

understanding by t_0 and t_1 the temporal bounds of the acceleration.

The deflection can therefore be reduced to an appropriate degree in all cases by the choice of the design quantities, and indeed again the oscillation period, which depends on N/P. The previously calculated

*) For the following calculation, see Schuler, l. c., p. 569.

deflection ψ_b for an east-to-west acceleration, calculated on the basis of the dimensions of the Anschütz apparatus to be given later, is much smaller than (namely, less than than 1/100 of) the deflection ψ_m for not too high a latitude, and is thus, in fact, not to be considered in practice.

The stability of the gyroscopic compass is further increased by the damping of the oscillations, which we could consider by a schematic additional term $+D\,d\psi/dt$ in equation (4). For the Anschütz apparatus, the actual friction is naturally kept to a minimum, so as not to restrict the required mobility of the rotor. An ingenious pendulum device provides artificial damping through an external moment about the vertical that is produced by the flow of air generated by the rapidly spinning rotor. The device is made in such a way that the moment is proportional to the elevation of the figure axis above the equilibrium position in the meridian, and therefore, according to equation (2a) with the neglect of ϑ'', proportional to the instantaneous precessional velocity, as is required for the damping. The Anschütz arrangement consequently has, moreover, a small deviation from the meridian in the equilibrium position for changes in latitude, and also, to a smaller degree, for motions of the ship; these deviations must be known in order to use the compass.

In the technical implementation, the horizontal equilibrium position of the axis that we ascribed schematically to a weight on the inner ring of the Cardanic suspension is achieved by hanging the rotor from a float. The rotor thus has the required three degrees of freedom, and friction is restricted to a minimum. The schematic arrangement is shown in Fig. 134. The rotor is borne in the housing s, from whose form the horizontal figure axis and the vertical plane of the rotor may be recognized. The housing s is rigidly bound to the float i through the connecting piece d. The float is in the shape of a ring, and can move in the housing k, which is filled with mercury. The connecting piece extends above the opening of the ring, and is rigidly fixed to the compass rose b. The center of gravity of the entire

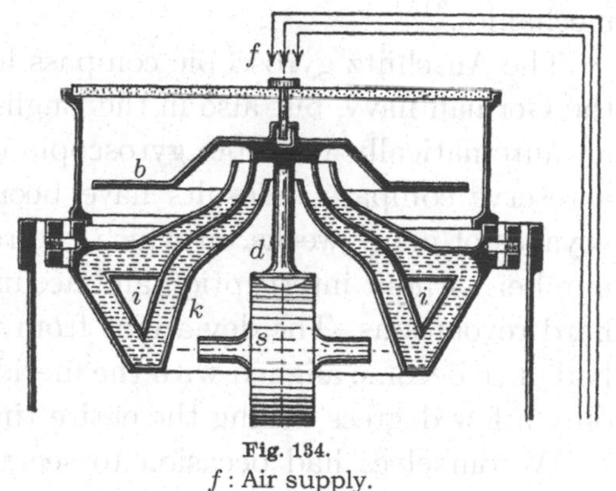

Fig. 134.
f: Air supply.

system lies beneath its metacenter, which is approximately the center of gravity of the immersed part, so that the equilibrium position is stable without rotation. Sufficient clearance is allowed for small vertical elevations of the entire system about the line of nodes.

The diameter of the rotor is 14,8 cm;*) its moment of inertia about the rotation axis is 136000 gr cm^2. The rotational velocity is held constant by an alternating current motor at 20000 rotations per minute. From these values, the impulse is calculated as $N = 28{\cdot}10^7$ gr cm^2 sec^{-1}. The gravitational moment P is $15 \cdot 10^6$ gr cm^2 sec^{-2}, corresponding to a counterweight of 1500 gr at a distance of 10 cm from the axis. The moment $N\omega\cos\varphi$ of the directional force of the compass is then, calculated for the latitude of Kiel, about 12000 gr cm^2 sec^{-2}; its apparent moment of inertia about the vertical is $\frac{N^2}{P} = 52 \cdot 10^8$ gr cm^2. Compared with this apparent moment of inertia of the horizontal precession, the actual moment of inertia B, which is of the order of magnitude of the moment of inertia about the rotation axis, can be neglected entirely. The resulting oscillation period is, as previously mentioned, about 70 minutes. All details of the construction, such as the electric drive, the lubrication of the ball bearings, etc., are naturally executed with the greatest care; the apparatus represents a masterpiece of precision mechanics.[315]

The Anschütz gyroscopic compass has been introduced not only in the German navy, but also in the English and Austrian. Each ship has an automatically readable gyroscopic compass in three locations and a reserve compass. Results have been published from experimental voyages of many weeks, in which the rotor ran at a constant rotation number without interruption and accumulated approximately one milliard revolutions. The deviations from the meridian, after the top axis had once become aligned with the meridian for two or three hours, were only a few degrees during the entire time.

We ourselves had occasion to see the operation of the gyroscopic compass installed by Mr. Anschütz in the Deutsches Museum in Munich, and to observe its certain alignment with the meridian in a relatively short time. Particularly striking was the stationarity of the compass rose in the presence of strong external vibrations. Certain undesirable deviating effects are caused, according to information from Mr. Anschütz, only by periodic vibrations in a fixed direction

*Taken from a construction drawing most kindly made available to us by Mr. Anschütz.

whose axis does not coincide with the northern direction, as could be transmitted, for example, by a rolling motion of the ship. Their effect has been disabled by the suspension of the entire system on a very long spring.[316]

On the whole, it is to be concluded that the gyroscopic compass offers an actually useful substitute for the magnetic compass. Its present very high cost, however, may prevent its immediate introduction into the merchant marine*) and restrict its primary domain of applicability to military ships, for which, according to the conditions described at the outset, a nonmagnetic directional indicator is of immediate and vital importance.

§8. Stability of the bicycle.

The most essential question for the construction of the modern bicycle, the question of the energy that the rider must expend for the attainment of a relatively large velocity, has been treated theoretically many times.**) Greater energy savings and easier maneuverability are the essential advantages of the bicycle over the tricycle. With these advantages, however, comes the disadvantage that the center of gravity is in unstable equilibrium in the upright position. In order to maintain this equilibrium position in the presence of external disturbances, a learning process is necessary for the rider, a process that has become easy, however, due to the special construction of the modern bicycle.

Can the gyroscopic effect of the rotating wheels also contribute significantly to the stabilization of the upright position at a sufficiently high velocity, in that, conventionally speaking, the rotation axes tend to maintain their direction in space? With consideration that the mass of the wheels is small with respect to the mass of the total system formed by the bicycle and the rider, one may doubt such an effect. And it is obviously not the intention of the manufacturer to strengthen this effect, since indeed his tendency, in the interest of energy savings, is to make all parts of the bicycle as light as possible, while an increase of the mass of the wheels would be advantageous for stabilization by the gyroscopic effect.

In any case, we must emphasize here too that the gyroscopic effect can come into play only if the system has a sufficient number of degrees

*) Cf. the discussion after the initially cited presentation by Anschütz to the Schiffbautechnische Gesellschaft, Jahrbuch X, p. 361.[317]

**) Cf. Encyklopädie der math. Wiss. Bd. IV, Nr. 9 (Walker, Spiel und Sport), p. 149.[318]

of freedom. For a "unicycle" (that is, a single rolling disk), stabilization is undoubtedly possible, as theoretical calculation has confirmed in agreement with experience. For a sufficiently high velocity, the rolling motion of such a disk in a vertical plane is stable.*) The motion can be conceived here as a translational motion of the center of gravity that is bound with a gyroscopic motion about the center of gravity; the latter is indeed known, from the previous investigations of this book, to be stable.

The formerly common "high wheeler" has the relatively greatest similarity to the simple disk. The large front wheel corresponds to the disk; the small rear wheel is added only to support the seat of the rider and to make steering possible. The rear wheel, however, diminishes the number of degrees of freedom of the system by one, and would thus make stabilization impossible if the turnable handlebars did not allow a rotation of the plane of the front wheel with respect to that of the rear wheel. For fixed handlebars, the entire system has only two degrees of freedom, the tipping about the horizontal track of the wheels and the forward motion bound to the rotation of the wheels; the possibility of stabilization by the gyroscopic effect is thus lost. The modern bicycle is different from the high wheeler only in its proportions; the two wheels are equally large, and the mass of the wheels is much smaller in relation to the total mass. Thus the influence of the gyroscopic effect is also weakened.

The third degree of freedom, the rotation about the steering axis, makes possible not only the gyroscopic effect, but also the assistance that the rider himself can apply to maintain the upright position of the bicycle, an assistance that the trained rider applies involuntarily. The original theory of the bicycle due to Rankine**) considered only this stabilization carried out by the rider. If, for example, the entire bicycle inclines to the right, the rider will rotate the front wheel to this same side, and thus force the bicycle to turn to the right. The centrifugal force at the center of gravity that arises from this turning has a moment about the track of the wheels that again uprights the plane of the bicycle.[320] In order to now avoid a tilt to the left, the rider must turn the handlebar to the left, and so on. Because this artificial stabilization also necessarily presupposes the third degree of freedom, the rotation about the steering axis, it is difficult to determine what

*) Cf. Carvallo, Journ. de l'École polyt. 2. Ser., 5. Heft, 1900.[319]

**) Theory of bicycle, Engineer 1869.

portion of the stabilization is due to the gyroscopic effect and what portion is due to the involuntary motions of the rider. Against the unconditional acceptance of Rankine's theory, the rider will object that he is completely unconscious of constantly guiding the handlebars, that he is indeed also able to ride safely without touching them, and that he guides the handlebars more to prevent a tipping of the front wheel than to stabilize his upright position. Indeed, he can also produce a gravitational moment that opposes a tilting by an involuntary lateral inclination of his body.

The extent to which stabilization can be achieved by small motions of the rider remains undecided; perhaps the matter could be decided by experiment. In any case, it is of interest to investigate the degree to which self-stabilization of the bicycle is possible without the motions of the rider, and the degree to which gyroscopic effects play a role. The process of stabilization is then the assistance discussed by Rankine, automatically taken over in part by the gyroscopic effects, and supported by the appropriate construction of the bicycle, which is yet to be discussed. The question of the extent to which the bicycle is stabilized without the participation of the rider, under the assumption that the rider is rigidly bound to the frame of the bicycle and does not hold the handlebars, is treated by Whipple*) and Carvallo.**) We will have to investigate, in the following, the degree to which gyroscopic effects take part in this stabilization. In this investigation we will naturally disregard all secondary circumstances (one-sided actuation by the pedals, deformability of the pneumatic tires and the consequent finite contact surface with the ground, friction in the steering axis, boring friction on the ground, etc.).

Whipple and Carvallo formulated the general Lagrange equations of the first and second kinds, the latter correspondingly modified because the bicycle is a "nonholonomic system," and specialized the equations to the case of small oscillations about upright motion in a straight line.[323] We hope to let the mechanical relations emerge more clearly in our derivation of the corresponding approximate equations by adding the gyroscopic effects and centrifugal forces generated by the motion to the forces that act on the system at rest, as was done for the previously discussed applications. It is sufficient, in order to retain

*) Quart. Journ. of Math., Nr. 120, 1899.[321]

**) Journal de l'École polytechnique, 2. Ser. 6. Heft 1901.[322]

terms of the first order in the oscillations, to apply the simplified expression (I) on p. 764 for the gyroscopic effect. If we neglect quadratic terms in the small oscillations, we note that the magnitudes of the angles under consideration lie fully within the bounds for which this approximation suffices for the precision that is required here.

The equations obtained in this manner agree with those of Whipple and Carvallo. It is to be concluded from these equations that the motion is naturally unstable for small velocities. For certain intermediate velocities, however, the motion becomes stable; the oscillations for these velocities can be represented in the form

$$Ae^{\lambda t},$$

where λ denotes a complex quantity with a negative real part. Whipple finds the range of stable velocities, under numerical assumptions that correspond better to a modern bicycle than those of Carvallo, to be approximately

$$16 \text{ kmh}^{-1} \text{ to } 20 \text{ kmh}^{-1},$$

and thus velocities that are easily attainable. For greater velocities the motion again becomes unstable, which could seem paradoxical, but it will be easy to explain this phenomenon from the way in which the individual components of the system are coupled. Practically, moreover, the latter instability is weak, and can be eliminated by almost imperceptible motions of the rider, even without touching the handlebars.

Our interest here is the contribution of the gyroscopic effect to the cited results. We will show something that was not pursued by the cited authors: the domain of complete stability would vanish in the absence of the gyroscopic effects, and thus the gyroscopic effects, in spite of their smallness, are indispensable for autonomous stabilization.[324]

The bicycle (Fig. 135) consists essentially of the frame, which carries the rear wheel in its plane, and the handlebars, whose axle carries the front wheel. Since the axis of the handlebars is guided in a tube that is fixed in the plane of the frame, the bicycle consists of two planar systems that are joined by a common axis of rotation. We assume that the rider is rigidly fixed to the frame. The rotation axis of the handlebars in the modern bicycle is inclined to the rear, and indeed is directed in such a way that its extension intersects the vertical line segment B_1S_1 from the contact point B_1 of the front wheel to its midpoint S_1 (Fig. 135);

the extension thus meets the ground in front of the contact point. As B o u r l e t*) emphasizes, the tilting of the front wheel is impeded by this arrangement of the steering axis if the rear wheel is guided or is imagined to be kept upright to a certain degree by a displacement of the center of gravity of the rider. A more detailed entry into the consequences of this not inessential arrangement is dispensable here, since its influence naturally appears of itself in our later analytic treatment.

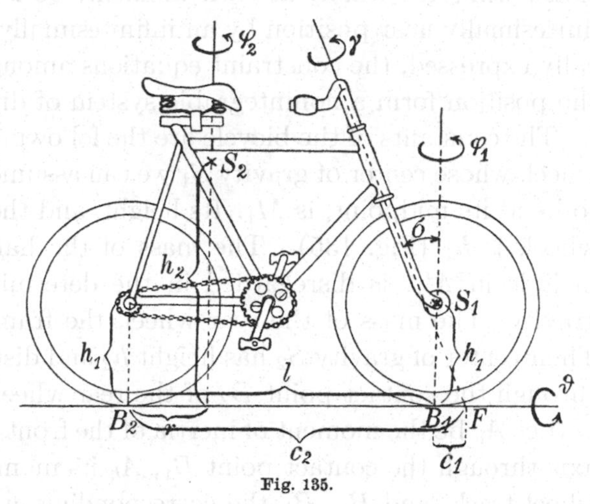

Fig. 135.

According to the given kinematic dependence between the front and rear wheels, we can arbitrarily prescribe, at any moment, the position of the plane of the frame (2 coordinates for the contact point, 2 coordinates for the position of the plane), and further prescribe the rotation of the plane of the front wheel with respect to that of the frame (1 coordinate). The position of the entire bicycle is then determined, since the condition is still to be added that both wheels must touch the ground; thus the bicycle must have, ignoring the cyclic coordinates that fix the positions of the wheels about their midpoints in their planes, a total of 5 degrees of freedom for its position. The possible motions of the bicycle are first an inclination of its frame, and then a rotation of the front wheel about the steering axis. Since both wheels are assumed to roll, the direction of motion of the front wheel is determined by the position of its plane. The motion of the steering axis is then given, and thus the motion of the plane of the rear wheel that is bound to the steering axis is also given. Only the velocity of the forward motion remains arbitrary. The bicycle thus has three degrees of freedom for its motion.

We recognize here the characteristic feature of nonholonomic systems,**) to which all rolling motions belong:[326] the bicycle can be led

*) In the attractive little book Nouveau traité des bicycles et bicyclettes, Paris 1898, p. 87 ff.[325]

**) Cf. H. H e r t z, Die Prinzipien der Mechanik, 1. Buch, Abschn. 4, Nr. 123–133.

into each of its ∞^5 possible positions by a sequence of allowable motions, while it cannot, at each moment, be led into each possible infinitesimally near position by an infinitesimally small motion. Analytically expressed, the constraint equations among the five coordinates of the position form a nonintegrable system of differential equations.

The constants of the bicycle are the following. The mass of the front wheel, whose center of gravity S_1 we can assume without essential error to be at its midpoint, is M_1. Its height, and therefore the radius of the wheel, is h_1 (Fig. 135). The mass of the handlebars, which we can include in M_1, is disregarded in the determination of the center of gravity. The mass of the rear wheel, the frame, and the rider is M_2. Their center of gravity S_2 has height h_2 and distance r from the vertical through the contact point B_2 of the rear wheel.

Let A_v be the moment of inertia of the front wheel about the vertical axis through the contact point B_1, A_h its moment of inertia about the wheel track, and B_v, B_h the corresponding quantities with respect to the contact point of the rear wheel for the system of the rear wheel, the frame, and the rider. Let B_{hv} be the product of inertia of this system in the plane of the wheel, likewise with respect to the contact point.

The steering axis has inclination σ with respect to the vertical; its intersection F with the ground lies at the distance c_1 in front of the contact point of the front wheel, and at the distance c_2 in front of the contact point of the rear wheel, so that $c_2 - c_1 = l$ is the distance between the two contact points.

Let ϑ_2 be the inclination of the plane of the rear wheel with respect to the vertical, where an inclination to the right-hand side of the rider is calculated as positive. Let ϑ_1 be the corresponding inclination of the front wheel, and γ the angle between the front and rear wheels, measured about the steering axis; γ is positive if the front wheel is inclined in the counterclockwise sense with respect to the rear wheel as seen from above. Let φ_1 be the angle of the front wheel with respect to the plane of the mean direction of travel, measured about the vertical in the same sense as γ, and φ_2 the corresponding angle for the rear wheel. The angles ϑ_1, ϑ_2, φ_1, φ_2, γ, are treated as very small angles, since only then do these definitions have a direct meaning. The kinematic formulas for the motion are to be derived under this restriction.*)

We imagine that the front wheel is brought into its inclined position ϑ_1 in two steps. It first has the inclination ϑ_2 as if it were rigidly bound

*) For an exact development of the kinematics of the bicycle, cf. Whipple and Carvallo, l. c.

to the rear wheel, and then rotates by γ about the steering axis. The latter rotation can be decomposed (see Fig. 135) into two components. The component about the track of the wheel is, with consideration of the previously defined sense of the rotation, $-\gamma \sin \sigma$. This is added to the inclination ϑ_2 of the rear wheel according to the theorems of small rotations, so that

$$\vartheta_1 = \vartheta_2 - \gamma \sin \sigma. \tag{1}$$

The vertical component of the rotation γ then gives

$$\varphi_1 - \varphi_2 = \gamma \cos \sigma. \tag{2}$$

Corresponding to the remark above, a nonholonomic constraint for the components of the motion is still to be added to these geometric constraints for the position coordinates. Let the mean velocity of the forward motion be u. We imagine that the position of the bicycle, as well as the velocity u and the rotation $\frac{d\varphi_1}{dt}$ of the front wheel about its contact point, are given. We obtain the desired constraint if we observe that these two data determine the motion of the footprint of the steering axis, and therefore also the motion of the rear wheel, which indeed follows the steering axis. The footprint F (Fig. 136) has velocity u in the direction of the track of the front wheel, and velocity

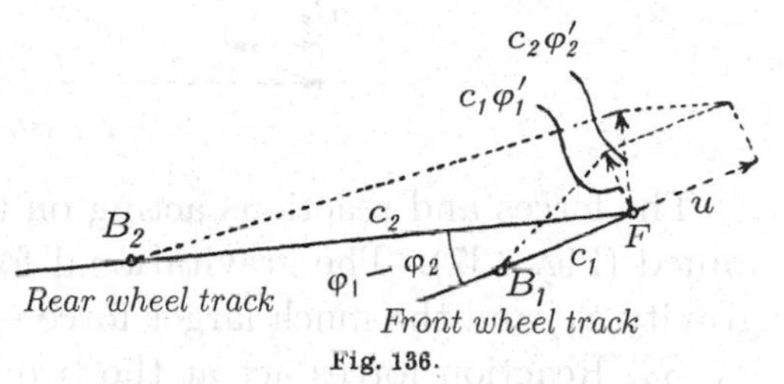

Fig. 136.

$c_1 \frac{d\varphi_1}{dt}$ perpendicular to this plane. We form the component of the motion perpendicular to the track of the rear wheel, which encloses the angle $\varphi_1 - \varphi_2 = \psi$ with the track of the front wheel. The desired component then becomes

$$c_1 \frac{d\varphi_1}{dt} \cos(\varphi_1 - \varphi_2) + u \sin(\varphi_1 - \varphi_2),$$

or, neglecting the quadratic terms,

$$c_1 \frac{d\varphi_1}{dt} + u(\varphi_1 - \varphi_2).$$

The lateral rotation $\frac{d\varphi_2}{dt}$ of the rear wheel would give the component

$$c_2 \frac{d\varphi_2}{dt},$$

and the desired nonholonomic equation is thus

$$c_2 \frac{d\varphi_2}{dt} = c_1 \frac{d\varphi_1}{dt} + u(\varphi_1 - \varphi_2). \tag{3}$$

Since c_1 is small in comparison with c_2, the constraint (3) states that the rear wheel must, in general, follow to the side to which the front wheel is turned with respect to the rear wheel. By consideration of the component of the motion of F along the track of the rear wheel, another constraint would follow for the velocity of the forward motion of the rear wheel, which would differ from u by terms of the second order. This constraint is inessential for our purpose.

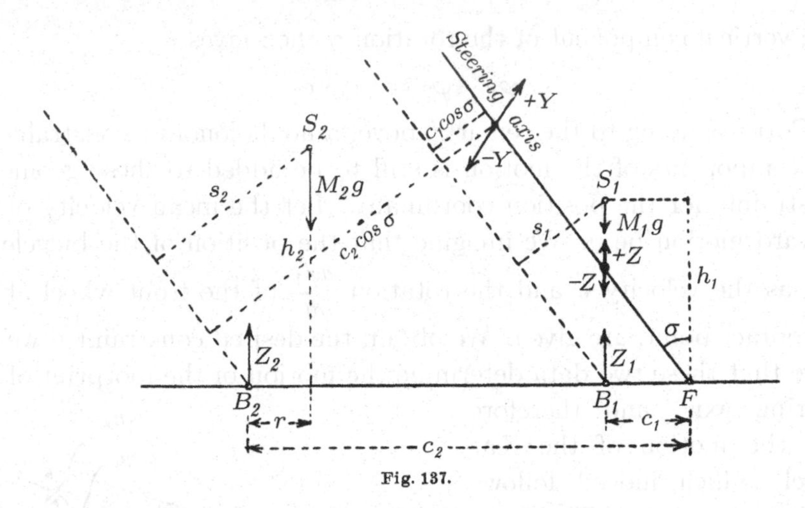

Fig. 137.

The forces and reactions acting on the bicycle are now to be determined (Fig. 137). The gravitational force $-M_1 g$ acts at the center of gravity S_1, and the much larger force $-M_2 g$ acts at the center of gravity S_2. Reaction forces act at the two contact points with the ground and at the steering axis that joins the front and rear wheels.

We first consider the vertical reactions. We need consider here only the reactions that maintain equilibrium with gravity in an upright straight-line motion, since these are finite quantities; their changes, in the case of small deviations from straight-line motion, are to be neglected as small quantities of higher order with respect to their original values.

Thus only the vertical forces that support the two parts of the bicycle in the case of equilibrium come into consideration. According to the general principles of mechanics, one must apply to the steering axis a reaction force and a reaction moment whose axis is perpendicular to the plane of the bicycle. The force and the moment can generally be composed, however, into a displaced single force, and we thus obtain the reactions as follows (Fig. 137).

First, as if the system were completely rigid, there is a force

$$Z_2 = M_2 g \frac{l-r}{l} \tag{4}$$

at the contact point of the rear wheel, and a force

$$Z_1 = M_1 g + M_2 g \frac{r}{l} \tag{5}$$

at the contact point of the front wheel. The component $M_2 g \frac{r}{l}$ in Z_1 is transmitted onto the steering axis by the bearing pressure of the frame; the two parts of the hinged system are thus in equilibrium if we add on the steering axis the reactions

$$\begin{aligned} -Z &= -M_2 g \frac{r}{l} \text{ on the front wheel,} \\ +Z &= M_2 g \frac{r}{l} \text{ on the frame,} \end{aligned} \tag{6}$$

each perpendicularly above the contact point of the front wheel. The height of their application point, according to Fig. 137, is $c_1 \operatorname{ctg} \sigma$.

We now turn to the horizontal reactions. First, a reaction $\pm Y$ acts on the steering axis perpendicular to the plane of the bicycle, and thus also perpendicular to Z; this reaction arises from the transmission of the lateral motion between the two parts of the system. Its magnitude, which is certainly of the first order, can naturally not be given statically, but rather depends on the instantaneous state of motion. The detailed determination of its point of application is inessential for our purpose.

Finally, the reactions on the steering axis in the direction of travel must still be introduced, which indeed actually transmit the drive of the rear wheel to the front wheel. These reactions are generally not present, however, in the case of uniform straight-line motion when rolling friction is neglected, and are only of the second order in the case of small deviations; since their lever arm about all axes in the vertical plane passing through the contact point is of the first order, their moments in the equations to be used below are of the third order, and thus do not come into consideration.

In the same manner, we need not consider that the contact point of the front wheel is displaced laterally by a rotation of the front wheel about the steering axis, as geometric intuition shows. For this displacement is of the first order, and its influence on the moment about the contact point is of the second order.

We must add here a kinematic remark regarding the position of the center of gravity.*) It is easy to show geometrically that for our assump-

*) Cf. Bourlet, l. c., p. 91.

tions about the positions of the steering axis and the center of gravity S_2 (namely $c_1 > 0$, $r > 0$), the height of the center of gravity S_2 is indeed unchanged in the first approximation, on the basis of symmetry, by a rotation of the front wheel when the rear wheel is vertical, but decreases when terms of the second order are calculated, so that the position $\gamma = 0$ is a maximum for the height of the center of gravity. This implies that in the gravitational potential, which corresponds for a simple lateral inclination to

$$\text{const} - \frac{g}{2}(M_1 h_1 \vartheta_1^2 + M_2 h_2 \vartheta_2^2),$$

terms with the factors $c_1 r g M_2 \gamma^2$ and $c_1 r g M_2 \vartheta \gamma$ will also appear. We can refrain from their precise calculation here, since the force terms that correspond to them will enter of themselves into the final equations by means of the reaction Z (equation (6)) on the steering axis.

In contrast, the kinetic terms that correspond to this descent do not come into consideration in our approximate equations of the first order, as is the case for all oscillations about an equilibrium position.

In addition to the already discussed forces that are present without forward motion, we must consider the apparent forces of the motion, the centrifugal forces and the gyroscopic effects. We divide the forward motion into the translation with velocity u in the mean direction of travel and the rotation of the wheels in their planes, whose angular velocity is $\frac{u}{h_1}$. The impulse of this rotation is the same for each of the two wheels, and is again denoted by N. The centrifugal force

$$M_2 u \frac{d\varphi_2}{dt}$$

is then applied to the center of gravity S_2, and the centrifugal force

$$M_1 u \frac{d\varphi_1}{dt}$$

is applied to the center of gravity S_1, perpendicular to the plane of the wheel.

The gyroscopic effects arise from the rotation that is bound to the forward motion; their magnitudes and senses are determined in the manner that was developed in detail for the case of the railroad carriage. First, a descent $\frac{d\vartheta}{dt}$ of the bicycle results in the vertically directed gyroscopic effects

$$-N\frac{d\vartheta_2}{dt} \text{ on the rear wheel,}$$

$$-N\frac{d\vartheta_1}{dt} \text{ on the front wheel.}$$

A descent to the right, for example, produces a moment about the vertical to the right (as seen by the rider).

Further, the gyroscopic effects

$$+N\frac{d\varphi_2}{dt} \text{ on the rear wheel,}$$

$$+N\frac{d\varphi_1}{dt} \text{ on the front wheel}$$

are to be added; these arise from the rotation $\frac{d\varphi}{dt}$ about the vertical, and act about the track of the wheels.

The latter two moments have the same direction as the moments of the centrifugal forces, and, since N is proportional to u, generally differ from those moments only by a constant factor. Since, however, the masses of the rider and the frame contribute to the centrifugal forces but only the masses of the wheels contribute to the gyroscopic effects, these components of the gyroscopic effects are inessential in comparison with the centrifugal forces, as in the case of the railroad carriage (cf. p. 775).

After this preparation, we can now state the impulse equations for the previously discussed degrees of freedom for the motion. For the stability question that interests us here, only the equations for the lateral inclination and the rotation about the steering axis come into consideration, while the simultaneously present small oscillations in the velocity of the forward motion have only an influence of higher order on these two coordinates.

The change of the impulse of the rear and front wheels about the track of the wheels is expressed kinetically in terms of mass and acceleration (A_{hv} is zero if we disregard the slanted position of the handlebars, whose mass is small and thus was not mentioned above) as

$$B_h\vartheta_2'' - B_{hv}\varphi_2'' \quad \text{for the rear wheel,}$$

$$A_h\vartheta_1'' \quad \text{for the front wheel.}$$

The sum of these terms is equal to sum of the moments acting about the track of the wheels, namely the gyroscopic effects about the track, the moments that arise from the centrifugal and the gravitational forces (lever arm h_1 or h_2), and finally the moments from the vertical reaction Z. We may form this sum algebraically instead of vectorially, since the consideration of a small angle between the two wheel tracks would give rise only to terms of the second order. We thus obtain

$$\begin{aligned} A_h\vartheta_1'' + B_h\vartheta_2'' - B_{hv}\varphi_2'' &= N(\varphi_1' + \varphi_2') + u(M_1h_1\varphi_1' + M_2h_2\varphi_2') \\ &+ M_2gh_2\vartheta_2 - Zc_1 \operatorname{ctg}\sigma \cdot \vartheta_2 + M_1gh_1\vartheta_1 + Zc_1 \operatorname{ctg}\sigma \cdot \vartheta_1. \end{aligned}$$

This equation becomes, after introduction of the force Z from (6),

$$(7)\quad \begin{cases} A_h\vartheta_1'' + B_h\vartheta_2'' - B_{hv}\varphi_2'' \\ \qquad = N(\varphi_1' + \varphi_2') + u(M_1h_1\varphi_1' + M_2h_2\varphi_2') \\ \qquad + g[M_1h_1\vartheta_1 + M_2h_2\vartheta_2 + M_2c_1 \operatorname{ctg}\sigma \frac{r}{l}(\vartheta_1 - \vartheta_2)]. \end{cases}$$

The last term

$$gM_2c_1 \operatorname{ctg}\sigma \frac{r}{l}(\vartheta_1 - \vartheta_2)$$

of equation (7), or, according to the kinematic equation (1),

$$-gM_2c_1 \cos\sigma \frac{r}{l}\gamma,$$

vanishes if the planes of the front and rear wheels coincide; this term corresponds to the previously remarked descent of the center of gravity S_2 caused by a rotation of the front wheel. Moreover, this term always vanishes if the steering axis passes through the contact point of the front wheel ($c_1 = 0$), since then, in the first approximation, the front wheel may turn freely about the steering axis, without influence of the rear wheel.

We have yet to form the impulse equations, for example, for the vertical axes, but prefer, for the sake of simplicity, to construct instead the equations for axes parallel to the steering axis, which are therefore inclined with respect to the vertical by the angle σ. We choose the axes through the two contact points, and first form, for the sake of greater transparency, the equations for the two parts of the bicycle separately, in which we must consider the reactions on the steering axis.

Of these reactions, only the moments of the forces $\pm Y$ perpendicular to the plane of the wheels and the vertical forces $\pm Z$ come into consideration; for the given axes, the sum of their moments is

$$\begin{aligned} &(Y + Z\vartheta_2)c_2 \cos\sigma && \text{on the rear wheel,} \\ -&(Y + Z\vartheta_1)c_1 \cos\sigma && \text{on the front wheel,} \end{aligned}$$

independent of their points of application. The reaction moment on the steering axis, in contrast, can remain out of consideration; the two parts of the bicycle are free to turn about the steering axis, so that this moment has, in any case, no component about the steering axis or the chosen axes parallel to it.

The impulse change for the front wheel in this direction, again calculated from the kinetic elements, is, in the notation of p. 868,

$$\cos\sigma A_v\varphi_1'' - \sin\sigma A_h\vartheta_1''.$$

The gyroscopic effect about this axis, composed of the gyroscopic effects given on pp. 872 and 873, is

$$-N(\vartheta_1' \cos\sigma + \varphi_1' \sin\sigma).$$

In addition, the moments of the centrifugal force, the gravitational force and the reactions Z and Y are to be formed. The lever arm of the centrifugal force is $s_1 = h_1 \sin\sigma$, and, correspondingly for the rear wheel, $s_2 = h_2 \sin\sigma + r\cos\sigma$ (cf. Fig. 137). We thus obtain the impulse equation

$$(8)\quad \left\{\begin{aligned} &\cos\sigma A_v \varphi_1'' - \sin\sigma A_h \vartheta_1'' \\ &= -N(\vartheta_1' \cos\sigma + \varphi_1' \sin\sigma) - M_1 s_1 u \varphi_1' \\ &\quad - gM_1 h_1 \sin\sigma \cdot \vartheta_1 - Zc_1 \cos\sigma \cdot \vartheta_1 - Yc_1 \cos\sigma. \end{aligned}\right.$$

For the case $c_1 = 0$ in which the steering axis passes through the contact point of the front wheel, the process of the motion would be completely represented by this equation and equation (7). The reactions Z and Y would fall out, since they would be applied on the steering axis that passes through the contact point of the front wheel, and could not influence the motion about the steering axis. The lateral motion of the rear wheel, on the other hand, would be given by the motion of the contact point of the front wheel. Thus the gyroscopic moments about the vertical acting on the rear wheel would be completely canceled by the reaction on the steering axis, and would not come into consideration for the entire system.

For the actual arrangement of the steering axis, however, $c_1 > 0$, and the motion of the front wheel is thus influenced by that of the rear wheel; a lateral rotation of the rear wheel is indeed carried over to the front wheel in the same sense, so that, for example, the gyroscopic effects of the two wheels strengthen each other.

Thus there is still an equation to be formed for the rear wheel, which is constructed in entirely the same manner as equation (8). The impulse change here is

$$\cos\sigma(B_v \varphi_2'' - B_{hv} \vartheta_2'') - \sin\sigma(-B_{hv} \varphi_2'' + B_h \vartheta_2''),$$

and the equation thus becomes[327]

$$(9)\quad \left\{\begin{aligned} &(\cos\sigma B_v + \sin\sigma B_{hv})\varphi_2'' - (\cos\sigma B_{hv} + \sin\sigma B_h)\vartheta_2'' \\ &= -N(\vartheta_2' \cos\sigma + \varphi_2' \sin\sigma) - M_2 s_2 u \varphi_2' \\ &\quad - gM_2(h_2 \sin\sigma + r\cos\sigma) \cdot \vartheta_2 + Zc_2 \cos\sigma \cdot \vartheta_2 + Yc_2 \cos\sigma. \end{aligned}\right.$$

We must eliminate the reaction Y from (8) and (9), and thus form the impulse change about the steering axis itself. We therefore calculate

$$c_2 \cdot (8) + c_1 \cdot (9),$$

or

$$(10)\quad \begin{cases} c_2[\cos\sigma A_v\varphi_1'' - \sin\sigma A_h\vartheta_1''] \\ + c_1[(\cos\sigma B_v + \sin\sigma B_{hv})\varphi_2'' - (\cos\sigma B_{hv} + \sin\sigma B_h)\vartheta_2''] \\ = - N[(c_2\vartheta_1' + c_1\vartheta_2')\cos\sigma + (c_2\varphi_1' + c_1\varphi_2')\sin\sigma] \\ - c_2M_1s_1\varphi_1'u - c_1M_2s_2\varphi_2'u \\ + g[-\sin\sigma(c_2M_1h_1\vartheta_1 + c_1M_2h_2\vartheta_2) + c_1\cos\sigma\frac{r}{l}M_2(c_1\vartheta_2 - c_2\vartheta_1)]. \end{cases}$$

Since, in any case, it is the difference rather than the absolute values of φ_1 and φ_2 that is decisive for the motion, we wish to use the kinematic equations (2) and (3) to introduce the angle γ between the planes of the front and rear wheels into (7) and (10). If we use the abbreviation

$$\gamma\cos\sigma = \varphi_1 - \varphi_2 = \psi,$$

then we have the equations

$$c_2\varphi_2' - c_1\varphi_1' = \psi u,$$
$$\varphi_2' - \varphi_1' = -\psi',$$

and therefore, since $c_2 - c_1 = l$,

$$(11)\quad \begin{cases} l\varphi_1' = \psi u + c_2\psi', \\ l\varphi_2' = \psi u + c_1\psi', \end{cases}$$

and thus

$$(11a)\quad \begin{cases} l\varphi_1'' = \psi' u + c_2\psi'', \\ l\varphi_2'' = \psi' u + c_1\psi''. \end{cases}$$

By means of this substitution, equations (7) and (10) become

$$\mathrm{I}\quad \begin{cases} [A_h\vartheta_1'' + B_h\vartheta_2'' - \frac{B_{hv}}{l}(c_1\psi'' + u\psi')] \\ \qquad - \frac{N}{l}[(c_2 + c_1)\psi' + 2u\psi] \\ \qquad - \frac{u}{l}[(M_1h_1c_2 + M_2h_2c_1)\psi' + (M_1h_1 + M_2h_2)u\psi] \\ \qquad - g[M_1h_1\vartheta_1 + M_2h_2\vartheta_2 - M_2c_1\frac{r}{l}\cdot\psi] = 0, \end{cases}$$

and

$$\mathrm{II}\quad \begin{cases} \left[\frac{c_2^2}{l}A_v\cos\sigma + \frac{c_1^2}{l}(B_v\cos\sigma + B_{hv}\sin\sigma)\right]\psi'' \\ \qquad - c_2\sin\sigma A_h\vartheta_1'' - c_1(B_{hv}\cos\sigma + B_h\sin\sigma)\vartheta_2'' \\ \qquad + \left[\frac{c_2}{l}A_v\cos\sigma + \frac{c_1}{l}(B_v\cos\sigma + B_{hv}\sin\sigma)\right]u\psi' \\ \qquad + N\Big[(c_2\vartheta_1' + c_1\vartheta_2')\cos\sigma + \frac{c_1^2 + c_2^2}{l}\sin\sigma\cdot\psi' \\ \qquad\qquad + (c_2 + c_1)\frac{u}{l}\sin\sigma\cdot\psi\Big] \\ \qquad + \left[(c_2^2M_1s_1 + c_1^2M_2s_2)\frac{u}{l}\psi' + (c_2M_1s_1 + c_1M_2s_2)\frac{u^2}{l}\psi\right] \\ \qquad + g\left[c_2M_1s_1\vartheta_1 + c_1M_2s_2\vartheta_2 - c_1c_2\frac{r}{l}\sin\sigma M_2\psi\right] = 0. \end{cases}$$

Equation II, the impulse equation about the steering axis, naturally contains, in essence, the effects that work toward a relative rotation of the two wheels. The second term of the third line represents the effect emphasized by Bourlet,*) a moment that opposes a rotation of the front wheel about the steering axis if c_1 is positive, and thus keeps the front wheel upright if one imagines that the rear wheel is guided in the upright position. Since, namely, the steering axis lies in front of the contact point of the front wheel, a pressure in the plane of the frame from the frame onto the front wheel obviously tends to turn the front wheel about its contact point, and indeed in such a way that the planes of the two wheels will approach each other.

The gyroscopic effects and the centrifugal effects in the fourth and fifth lines contain terms with ψ' and ψ. The latter terms depend on the kinematic constraint that the presence of a rotation ψ requires a curvature of the trajectory of the bicycle, which, in its turn, must again be accompanied by deviating inertial effects. We point out here that the terms with ϑ_1' and ϑ_2' in the fourth line arise only from the gyroscopic effects.

In the case of a lateral inclination of the bicycle, the front wheel will be turned by gravity about the steering axis toward the side of the inclination, since indeed the center of gravity lies in front of the slanted steering axis. The first two terms of the last line correspond to this effect. The last term, in contrast, which contains the factor $-\psi$, corresponds to the previously mentioned descent of the center of gravity that is caused by a turning of the front wheel if c_1 is positive. The consequence of this kinematic relation must be, in fact, that an initial turning will be further increased by the gravitational force.

Equations I and II contain only the coordinates ϑ_1, ϑ_2, ψ. We add the kinematic equation (1)

$$\text{III} \qquad \vartheta_1 - \vartheta_2 = -\gamma \sin\sigma = -\psi \operatorname{tg}\sigma$$

as the third linear equation among these variables.**)

In order to investigate the stability of the system, we must set

$$(12) \qquad \vartheta_1 = a \cdot e^{\lambda t}; \quad \vartheta_2 = b \cdot e^{\lambda t}; \quad \psi = c \cdot e^{\lambda t}.$$

We then obtain, in a well-known manner, an algebraic equation for λ, and indeed an equation of the fourth degree, since the sum of the orders

*) Bourlet, l. c., p. 90.

**) Our equations I, II, III are linear combinations of those obtained by Whipple (p. 323), as well as those obtained by Carvallo (p. 100).

of the differential equations I and II is 4, while III contains no differential quotients. Thus two oscillations of the system exist; they are stable if all four roots are complex with nonpositive real parts, or, in the case of two real roots, if these two roots are negative. If we insert (12) into I, II, III, then we obtain, after cancellation of the factor $e^{\lambda t}$, three homogeneous linear equations in a, b, c. The determinant Δ of their coefficients is to be set to zero; this determinant Δ is

$$\Delta = \begin{vmatrix} A_h\lambda^2 \\ -gM_1h_1 & B_h\lambda^2 \\ -gM_2h_2 & -\frac{B_{hv}}{l}c_1\lambda^2 - B_{hv}\frac{u}{l}\lambda \\ -\left[N\frac{c_2+c_1}{l} + \frac{c_2M_1h_1+c_1M_2h_2}{l}u\right]\lambda \\ -\left[N\frac{2u}{l} + \frac{M_1h_1+M_2h_2}{l}u^2\right] \\ +gM_2c_1\frac{r}{l} \\ \\ -c_2\sin\sigma\cdot A_h\lambda^2 \\ +Nc_2\cos\sigma\cdot\lambda \\ +gc_2M_1s_1 & -[c_1\sin\sigma\cdot B_h + c_1\cos\sigma\cdot B_{hv}]\lambda^2 \\ +Nc_1\cos\sigma\cdot\lambda \\ +gc_1M_2s_2 & \left[c_2^2\cos\sigma\cdot\frac{A_v}{l} + c_1^2\frac{(\cos\sigma\cdot B_v + \sin\sigma\cdot B_{hv})}{l}\right]\lambda^2 \\ \left[c_2\frac{u}{l}\cos\sigma\cdot A_v + c_1\frac{u}{l}(\cos\sigma\cdot B_v + \sin\sigma\cdot B_{hv})\right]\lambda \\ \left[N\frac{c_1^2+c_2^2}{l}\sin\sigma + \frac{u}{l}(c_2^2M_1s_1 + c_1^2M_2s_2)\right]\lambda \\ +N\frac{u}{l}(c_1+c_2)\sin\sigma + \frac{u^2}{l}(c_2M_1s_1 + c_1M_2s_2) \\ -gc_1c_2M_2\frac{r}{l}\sin\sigma \\ \\ 1 & -1 & \operatorname{tg}\sigma \end{vmatrix}$$

If one considers that N is proportional to u and understands by α, β, γ, δ, ε constants that are independent of u, then the calculation of this determinant would give an equation of the form

$$\alpha\lambda^4 + \beta u\lambda^3 + (\gamma_1 + \gamma_2 u^2)\lambda^2 + (\delta_1 u + \delta_2 u^3)\lambda + (\varepsilon_1 + \varepsilon_2 u^2) = 0.$$

In order to fulfill the condition for stability, which states that this equation may have no root with a positive real part, all the coefficients of this equation, as is easy to see, must be positive. On the other hand, however, stability is not ensured by positive values of the coefficients; the further formation of a discriminant is required.

We do not wish to carry out this extensive calculation here, but rather refer, for the present, to the results obtained by Whipple, which correspond to the dimensions of a modern bicycle.

For small u, in any case, the system is unstable because of its weight. For small u, correspondingly, the coefficients

$$\gamma_1,\ \delta_1 \text{ are negative,}$$

and ε_1, in contrast, is positive, as are α and β. The coefficients of the higher powers of u,

$$\gamma_2 \text{ and } \delta_2, \text{ are positive,}$$

while ε_2, in contrast, is negative, but is small in absolute value compared with the other coefficients. The coefficients of λ and λ^2 thus become positive with increasing velocity; the constant term, in contrast, will ultimately become again negative.

And indeed, the coefficient $(\delta_1 u + \delta_2 u^3)$ of λ will first become positive for the approximate velocity

$$u_1 = 12 \text{ km/h},$$

and the coefficient $(\gamma_1 + \gamma_2 u^2)$ of λ^2 will then also become positive for the approximate velocity

$$u_2 = 14 \text{ km/h}.$$

Finally, the last coefficient $(\varepsilon_1 + \varepsilon_2 u^2)$ will become negative for

$$u_3 = 20 \text{ km/h}.$$

Stability is possible only between the limits u_2 and u_3, the domain in which all the coefficients are positive. A detailed discussion shows that complete stabilization actually occurs in the domain

$$u_4 = 16 \text{ km/h to } u_3 = 20 \text{ km/h}.$$

The calculations of Carvallo, which are carried out for an older model, give qualitatively the same but somewhat lower values for all these bounds.

We supplement these results by the proof that complete stabilization would not be possible without the gyroscopic effects. For this purpose, we calculate the coefficient $(\delta_1 u + \delta_2 u^3)$ of λ from the determinant Δ. If we denote the total gravitational moment $M_1 h_1 + M_2 h_2$ by Mh, then this coefficient becomes

$$\begin{aligned}
&g(-M_1h_1c_1 + M_2h_2c_2)\sin\sigma\cdot N\\
&- gMh[c_2\cos\sigma A_v + c_1(\cos\sigma B_v + \sin\sigma B_{hv})]\frac{u}{l}\\
&- gMh(c_1^2 + c_2^2)\sin\sigma\frac{N}{l}\\
&- gMh[c_2^2 M_1 s_1 + c_1^2 M_2 s_2]\frac{u}{l}\\
&+ g(c_2 M_1 s_1 + c_1 M_2 s_2)\Big[(c_1 + c_2)\frac{N}{l} + (M_1h_1c_2 + M_2h_2c_1)\frac{u}{l} + B_{hv}\frac{u}{l}\Big]\\
&+ (c_2 + c_1)\cos\sigma\cdot\frac{N}{l}[2Nu + Mhu^2 - gM_2c_1r],
\end{aligned}$$

and reduces to

$$\begin{aligned}
&- gMh\cos\sigma(c_2A_v + c_1B_v)\frac{u}{l}\\
&+ gB_{hv}\Big(-M_1h_1\sin\sigma + M_2r\frac{c_1}{l}\cos\sigma\Big)u\\
&- gM_1h_1M_2h_2 l\sin\sigma\cdot u - gM_1M_2h_1c_1r\cos\sigma\cdot u\\
&+ \frac{c_2 + c_1}{l}\cos\sigma N[2Nu + Mhu^2].
\end{aligned}\tag{13}$$

In this expression, the last term contains the factor u^3, since N is proportional to u; the other terms contain only the factor u. Of these others, the negative terms far overwhelm the positive term, since the latter contains the two small factors c_1 and r; the entire coefficient will thus be negative for small velocity u. It always remains negative, and thus the upright motion remains unstable if the gyroscopic effects remain unconsidered, and therefore $N = 0$ is assumed (that is, since the rotational velocity is proportional to u, if the moment of inertia of the wheels about their rotation axes is neglected). Because of the last term that arises from the gyroscopic effects, the term that contains the factor u^3, the coefficient will become positive for a sufficiently large velocity. (The order of magnitude of the velocity ranges distinguished here as small and sufficiently large can be seen from the numerical values of Whipple given above. The boundary between the two is formed by the value $u_1 = 12$ km/h.)

The stability of the bicycle found by Whipple for velocities of 16–20 km/h is thus made possible only by the gyroscopic effects of the rotating wheels.[328]

The following remarks should clarify how the gyroscopic effects come into operation. The factor N in the last term

$$(14) \qquad \frac{c_1 + c_2}{l} \cos \sigma N (2Nu + Mhu^2)$$

arises from the gyroscopic effect $-N\frac{d\vartheta}{dt}$ about the vertical axis. This effect, which rotates the front wheel to the same side as a lateral inclination of the bicycle and thus forces the bicycle to turn to this side, is therefore required for stabilization. In contrast, the quantity $2Nu$ in the term

$$2Nu + Mhu^2$$

arises from the uprighting gyroscopic effect $N\frac{d\varphi}{dt}$; this quantity is only added to the much greater and equally directed moment Mhu^2 of the centrifugal force of the entire system, and is therefore inessential.

The stabilizing effect of the rotation thus depends on the fact that when the bicycle inclines laterally, it is forced by the gyroscopic effect of the front wheel to turn, and thus the centrifugal force that uprights the bicycle occurs. The actual stabilizing force that overcomes the gravitational force is the centrifugal force; the gyroscopic effect plays the role of a release. Moreover, the capacity for stabilization diminishes with increasing inclination of the steering axis with respect to the vertical, because of the factor $\cos \sigma$ in (14).

Indeed, the gravitational force on the front wheel and the reaction Z from the rear wheel also act to turn the front wheel in the direction of a lateral inclination of the system, and thus awaken a centrifugal force. Nevertheless, according to the preceding, this effect is incapable of stabilizing the system completely. The gyroscopic effect is the only force that is proportional to $\frac{d\vartheta}{dt}$; the gravitational force is proportional to ϑ itself, and thus the gyroscopically induced turning follows the lateral inclination more rapidly than the gravitational effect. The gyroscopic effect, in fact, is displaced by only a quarter oscillation period with respect to the descent, while the gravitational effect is displaced by half an oscillation.

In order to understand the basis of the instability that occurs for large velocity, we consider the λ-free term $\varepsilon_1 + \varepsilon_2 u^2$ in the equation $\Delta(\lambda) = 0$. This expression, as is easily seen from the determinant Δ, contains terms with the factor g^2 that do not contain u. The sum of these terms, which will be positive, is[329]

$$(15) \quad g^2 M_1 M_2 h_1 \sin \sigma (\operatorname{tg} \sigma\, h_2 l - c_1 r) + g^2 M_2^2 c_1 r \left(h_2 \sin \sigma - \frac{c_1 r}{l} \cos \sigma \right).$$

The two negative terms in (15) are essentially smaller, because of the factors $c_1 r$ and $c_1^2 r^2$, than the positive terms. In addition, the terms

$$-gMh\Big(\frac{c_2+c_1}{l}Nu\sin\sigma+(c_2M_1s_1+c_1M_2s_2)\frac{u^2}{l}\Big)$$

$$+g(c_2M_1s_1+c_1M_2s_2)\Big[N\frac{2u}{l}+Mh\frac{u^2}{l}\Big]$$

or

$$gNu\Big[(M_1h_1-M_2h_2)\sin\sigma+\frac{2c_1}{l}M_2r\cos\sigma\Big] \tag{16}$$

with the factor u^2 appear.

Here the negative term $-gNuM_2h_2\sin\sigma$ that contains the weight of the rider dominates, while the last term is insignificant because of the small factors c_1 and r. Moreover, the terms (16) overwhelm the terms (15) for large u; thus the entire coefficient will become negative for large velocity, and the motion will become unstable.

We now trace the intrinsic origin of the terms (16); they correspond to the terms with the factor $u^2\psi$ in the differential equation II, and arise through the introduction into (10) of the kinematic equations

$$\psi=\varphi_1-\varphi_2, \tag{2}$$

$$c_2\varphi_2'=c_1\varphi_1'+u\psi. \tag{3}$$

These equations, however, express the intuitive fact that in the case of a relative rotation between the planes of the two wheels, the plane of the rear wheel, disregarding external forces, constantly approaches the plane of the front wheel during the motion, since it always passes through an axis that is fixed in the plane of the front wheel. If one imagines that the front wheel is guided, then the track of the rear wheel tends asymptotically to that of the front wheel (one can therefore denote it as a tractrix). The approach is very rapid for large velocity, so that for $u=\infty$ it is to be generally concluded from the kinematic equation (3), if the two oscillations of the front wheel do not also occur very rapidly, that

$$\psi=\varphi_1-\varphi_2=0.$$

The bicycle then behaves, however, as if its two parts were rigidly bound.

The stabilization by the gyroscopic effect depends, however, just on the relative mobility of the two wheels. As soon as they are rigidly bound, we can compare the entire system to a simple top that no longer has three degrees of freedom, but rather, because of the double contact with the surface of the Earth, only two, and can apply the general principle stated in §1, according to which any possibility of stabilization by the gyroscopic effect ends with the elimination of one degree of freedom. This explains the initially paradoxical phenomenon that the capability of rotation for the stabilization of the bicycle is lost for large velocities, while large rotational velocities are favorable for the stabilization of the free top. The unstable motion of the bicycle for large velocity will be overlaid with stable oscillations, which one can compare with the nutations of the free top. Because of their short period, our preceding argument does not apply to them.

It is still to be mentioned, moreover, that the motion remains stable for an arbitrarily large velocity when the steering axis is nearly vertical, since the last coefficient of the equation $\Delta(\lambda) = 0$ remains positive for small values of σ. In this case, the two eigenoscillations are sufficiently rapid to preserve stability in spite of the apparently rigid coupling of the front and rear wheels. The capability of the gyroscopic effect for stabilization, however, decreases in accordance with general principles as the inclination of the steering axis increases, since the two noncyclic degrees of freedom, the rotations about the wheel track and the steering axis, now approach one another.

We wish to briefly compare the obtained results with experience. It is to be emphasized once more that the assumption that the rider is rigidly bound to the bicycle, which lies at the basis of the entire discussion, cannot be realized practically, since the rider can always influence the stability of the bicycle through involuntary, almost imperceptibly small motions. The assistance that he can give here is of two kinds. First, he can turn the front wheel slightly about the steering axis at an appropriate point of time, and thus influence the motion centrifugally. The required rotations, like those produced automatically by the gyroscopic effect, are very small. In addition, the rider can produce an opposing gravitational moment that uprights the falling bicycle by a lateral inclination of his body. In hands-free riding, the rider abstains from the first effect and has only gravity at his disposal.

Experience confirms the existence of a lower bound of velocity beneath which hands-free riding becomes impossible, apparently because the assistance that lateral inclinations of the body can provide is insufficient as the self-stabilization of the bicycle ceases. An upper bound, in contrast, is not observed in practice. It may well be, corresponding to the detailed numerical calculation of Whipple, that the instability of the bicycle for attainably large velocities is very slight, so that imperceptibly weak inclinations of the body suffice to maintain stability.

Even if, according to the preceding, the self-stabilization of the bicycle has not been directly proven as necessary because of the small assistance that the trained rider can give, and even if the question of stability is less important for technical construction than the question of the greatest possible energy savings, it is hardly to be denied that the gyroscopic effect contributes to the maintenance of upright equilibrium in travel, and it contributes, we would say, in a particularly intelligent manner; it is the gyroscopic effect, thanks to its phase, that first senses a falling of the bicycle, and then presses the much stronger but somewhat slow centrifugal effect into the service of stability.

§9. On alleged and actual gyroscopic effects for the Laval turbine.

The ingenious construction of the Swedish engineer de Laval has provided technology with steam turbines of extraordinary speed and a particularly high degree of effectiveness.[330] The primary interest of the construction naturally lies in the domains of thermodynamics and hydrodynamics, in the direct actuation of the turbine by rapidly flowing steam and the ingenious form of the Laval nozzle. These new ideas in turbine construction have been consequential not only for technical practice, but also for technical science; research in the flow and pressure relations in a Laval nozzle now makes up an important special field in technical thermodynamics.*)

What interests us here, however, is a pure dynamic problem that de Laval solved immediately. No bearings and no shaft could withstand the high angular velocity of the Laval turbine (20 000 to 30 000 revolutions per minute for smaller implementations); any eccentricity of the bearings and any bending of the shaft would lead, due to the enormous centrifugal forces, to the destruction of the entire assembly. Laval's

*) Cf. the report of L. Prandtl in the Encyclopädie der mathem. Wiss., Bd. V, Art. 4.[331]

solution to this problem is quite remarkable: he made the shaft flexible (about the thickness of a finger), and thus achieved a kind of self-centering of the shaft that has often been used since for particularly high angular velocities. The shaft of the gyroscopic compass, with its thickness of a few millimeters and its 20 000 revolutions per minute (cf. p. 862), is also effectively flexible; one can hear an indication of the critical whirling (see below) as the shaft passes through the speed of about 10 000 revolutions per minute when the compass is actuated. The result achieved by a flexible shaft can also be obtained by compliant bearings. This means of construction has also been frequently used in technology since the time of Laval.

When the Laval shaft became known, it was believed that its astonishing behavior should be attributed to a gyroscopic effect. Just as the top allegedly adjusts itself to a uniform rotation about a stable axis, so should the Laval turbine wheel seek a stable form of rotation. The comparison is erroneous on both sides. The rotation axis of the top generally changes, in that it describes a precession; if the turbine wheel tends to a uniform rotation as the angular velocity increases, this depends on other and fundamentally much simpler dynamic principles than those of the motion of the top.*)

The behavior of a flexible shaft in rapid rotation will therefore be discussed not as a true application, but rather as an example of a false application of the theory of the top.**) It will be most easily understandable if the problem is posed in the context of the technically important theory of resonance (No. 1). That this theory suffices to explain the observed phenomena of the initial whirling and subsequent straightening of the shaft can be proven numerically by a model (No. 2). The gyroscopic effect comes into play only secondarily, if one abandons the simplest arrangement and places the turbine wheel not in the middle of the shaft, but rather, for example, at its free end, or if more than one turbine wheel is placed on the same axis (No. 3). To this extent, this section is related to the preceding treatment of the bicycle, for which the gyroscopic effect was also more of a secondary nature.

*) Cf. A. F ö p p l, Civil-Ing., p. 333, 1895.[332]

**) The relation to the theory of the top is naturally presented with complete correctness in the classic textbook of A . S t o d o l a, Die Dampfturbine, Berlin 1905, 3rd ed.[333] For the theory of the critical speed, cf. Nr. 62–70 of his book. Further references to the literature are found in the Encyclopädie der mathem. Wiss. IV 27, Art. Kárman, Nr. 13 b.

1. The self-positioning thin shaft.

We consider the following idealized model of the Laval arrangement (cf. Figs. 138 a and b).

The thin shaft (length $2l$) is supported in fixed bearings at its ends P and Q; it carries the turbine wheel R (mass M) at its center. Let S be the center of gravity of the wheel; S is assumed here to be exactly on the midline of the shaft (or is assumed to be displaced very little from the midline of the shaft, as the midline of the shaft is displaced very little from the midline PQ of the bearings). Let O be the midpoint

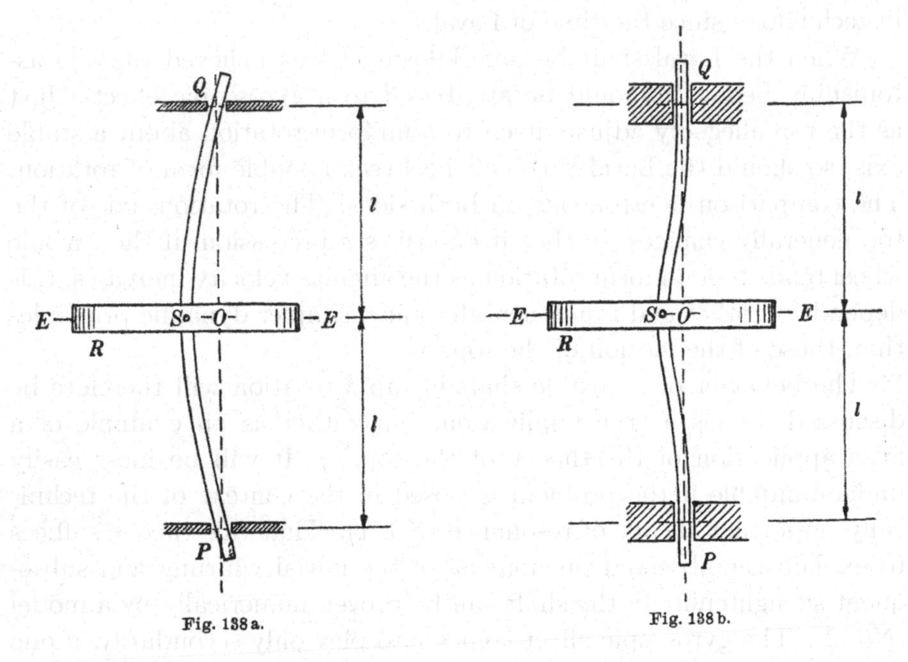

Fig. 138 a. Fig. 138 b.

of the line that connects the bearings P and Q. Since the shaft at rest will never be exactly straight, we must calculate with a certain original distance e between the points O and S. The distance e is transformed during the motion into the distance r. In order to eliminate the inessential effect of gravity, we place the line PQ most conveniently along the vertical. We can disregard the mass of the shaft in relation to the mass M of the wheel; we imagine the mass M as concentrated at the point S. We can also neglect, as is usual in the theory of beam bending, the difference between the actual length of the bent shaft and the length $2l$ of the line between its endpoints.

We thus have the following extremely simple image, for which we need speak further only of the midplane EE between points P and Q.

Point S of the mass M is held fixed by symmetry in the plane of Fig. 139, and rotates in this plane with the given angular velocity ω about the fixed point O. It is subject to the influence of an elastic force F that acts toward O, and seeks to bring it to the distance e from O. We seek the actual distance r of the point S from O for each given angular velocity ω.

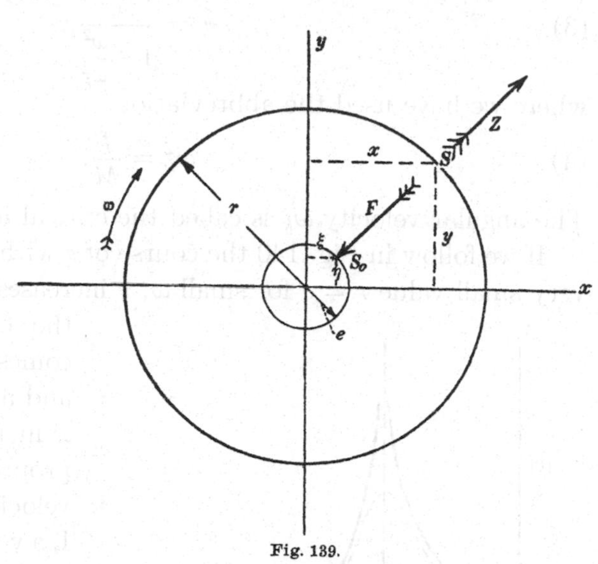

Fig. 139.

The force F has its origin in the elasticity of the shaft; with the restriction to small deformations $r - e$, one can set it proportional to this deformation, and thus assume

$$(1) \quad F = f(r - e).$$

The coefficient f depends on the particular bearings of the shaft. The two extreme possibilities, which are represented in Figs. 138a and 138b, are: a) *free bearings*: the bearings allow free mobility about the assumed fixed endpoints P, Q; b) *fixed bearings*: the bearings fix not only the endpoints P, Q, but also the direction of the shaft. The values of f corresponding to these two limiting cases are

$$\text{a)} \ f = \frac{6EJ}{l^3}, \qquad \text{b)} \ f = \frac{24EJ}{l^3},$$

where E and J signify the modulus of elasticity and the moment of inertia of the cross section (imagined, for example, as circular) of the shaft about one of its diameters. In reality, the mobility about the ends is not completely unimpeded by the free bearings, and the restraint of the fixed bearings is not complete because of a small compliance. There will thus be an intermediate state, with a value of f between the bounds

$$(2) \qquad 6 < \frac{l^3 f}{EJ} < 24.$$

In the state of uniform rotation with angular velocity ω, the elastic force F is in equilibrium with the centrifugal force

$$Z = Mr\omega^2.$$

Thus

$$f(r-e) = Mr\omega^2,$$

or

$$r = \frac{e}{1 - \frac{\omega^2}{\omega_k^2}}, \tag{3}$$

where we have used the abbreviation

$$\omega_k^2 = \frac{f}{M}. \tag{4}$$

The angular velocity ω_k is called the critical angular velocity.

If we follow in Fig. 140 the course of r with increasing ω, we see the very small value $r = e$ for small ω; r increases rapidly if ω approaches the critical value ω_k; r becomes negative if ω_k is crossed, and approaches the value 0 as ω increases further. As an appropriate value for the angular velocity of the steam engine, Laval chooses $\omega = 7\omega_k$. According to equation (3), this corresponds to a deviation r from exact centering that amounts only to the 48th part of the originally small eccentricity. In the figure, $|r|$ is plotted as the ordinate; that is, the sign has been disregarded. The negative sign of r for $\omega > \omega_k$ obviously signifies that the shaft is bent to the opposite side as for $\omega < \omega_k$. The figure is completely identical with the "fundamental curve" (equation (18)) in Fig. 125 of §5, which was designated there as the simplest type of resonance curve; the square of the current ordinate was plotted there. Complete centering or self-positioning of the turbine wheel will therefore be approached asymptotically. In this final state, the whirling effect is no longer transmitted to the bearings; the uniformity of the rotation is ideal. In the vicinity of the critical angular velocity, in contrast, the bearings and the shaft are seriously endangered. In the actual implementation, arresters that prevent dangerous bending of the shaft are used.

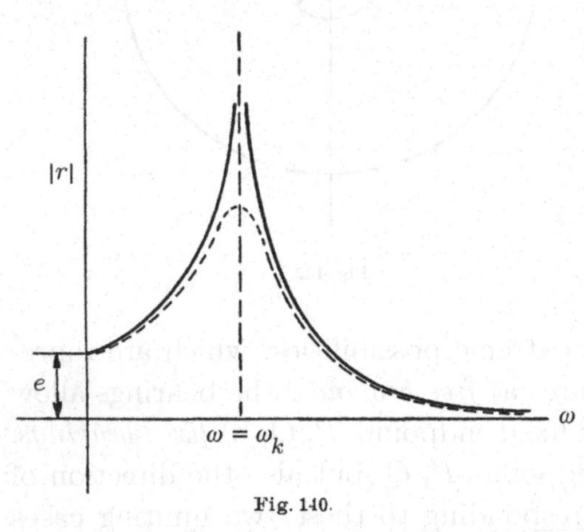

Fig. 140.

For a small model (cf. below), it suffices to hold the shaft between two fingers in order that the critical state can be crossed without danger.

It is self-evident that the preceding consideration gives only the final state corresponding to each ω. In reality, the passage from one value of ω to another will be accompanied by oscillations about the final state, which, however, are damped temporally. We will return to these oscillations directly. They are important only for the proof of the stability*) of the uniform rotation.

In order to emphasize the relation with the resonance principle, a second treatment of the same problem in coordinates is recommended. Let the rectangular coordinates of S with respect to a fixed coordinate frame through O (Fig. 139) be x, y; let ξ, η be the similarly measured coordinates of an auxiliary point S_0, the point to which S tends due to the elasticity, so that S_0 has distance e from O and lies on the same radius vector as S. We decompose the elastic force F into rectangular components F_x, F_y. According to the assumption (1), and with consideration of the sense of this force,

$$F_x = -f(x - \xi), \quad F_y = -f(y - \eta). \tag{5}$$

We must now disregard the centrifugal force Z, since we wish to express the accelerations in our equations directly. These equations are now

$$M\frac{d^2x}{dt^2} = -F_x, \quad M\frac{d^2y}{dt^2} = -F_y,$$

or, with consideration of equations (4) and (5),

$$\frac{d^2x}{dt^2} + \omega_k^2 x = \omega_k^2 \xi, \quad \frac{d^2y}{dt^2} + \omega_k^2 y = \omega_k^2 \eta. \tag{6}$$

These equations of motion for x and y may easily be integrated if the right-hand sides (that is, the coordinates ξ, η of our auxiliary point S_0) are known. Strictly speaking, however, this is not the case. We remark, however, that ξ and η are very small quantities, in that indeed $\xi^2 + \eta^2 = e^2$. Thus it is permitted to calculate them approximately, in that S_0 is assumed to rotate uniformly. We therefore write, if we denote the angular velocity by ω,

$$\xi = e\cos\omega t, \quad \eta = e\sin\omega t. \tag{7}$$

*) Cf. S t o d o l a, l. c., Nr. 103.

Equations (6) then have the form of the simplest oscillation equations. The two degrees of freedom x and y are independent of one another; both have the eigenfrequency ω_k, as is shown by the left-hand sides of (6). The terms on the right-hand sides act as external forces in the directions of x and y with period ω and very small amplitude $\omega_k^2 e$. Such forces produce forced oscillations of the x- and y-coordinates with the same period ω, and, in general, with very small amplitude, except when their period approximately coincides with the eigenperiod of the oscillating system. In the latter case, small forces suffice to cause significant deflections; we have the well-known phenomenon of resonance.

It follows from (6) and (7), if one makes the assumption $x = r\cos\omega t$, $y = r\sin\omega t$, that

$$r = \frac{e\omega_k^2}{\omega_k^2 - \omega^2},$$

and the expression for the forced oscillation is therefore

$$\left.\begin{matrix} x \\ y \end{matrix}\right\} = \frac{e}{1 - \dfrac{\omega^2}{\omega_k^2}} \begin{matrix} \cos \\ \sin \end{matrix}\, \omega t, \tag{8}$$

in agreement with (3). The same process that is represented in (3) as a "circular" oscillation is decomposed in (8) into two "linearly polarized" oscillations.

This is, however, only a particular form of the motion. We obtain the general integral of (6) if we add the free oscillations. They follow, by setting the left-hand sides of (6) to zero, as

$$\left.\begin{matrix} x \\ y \end{matrix}\right\} = A\cos\omega_k t + B\sin\omega_k t. \tag{8'}$$

In reality, the periodic bending of the shaft is opposed by a considerable material resistance that we have not considered in (6). This resistance produces a rapid diminishment of the free oscillations in time, and a strong reduction of the actual amplitude of the forced oscillations compared with that calculated in (8).

We thus confirm the result contained in equation (3) and Fig. 140, and supplement it in the following manner: the uniform rotation with period ω (the forced oscillation) is, in general, overlaid with oscillations of period ω_k that are generated anew for each change of velocity, but are soon damped because of internal resistance. The form of the amplitude in Fig. 140 is to be changed, with consideration of this resistance, in the sense of the dotted curve. *The critical rotation number is proven as the eigenfrequency of the shaft* (and indeed as its "fundamental frequency"; the higher oscillations do not come into consideration because of their rapidity) *and the entire process as a typical resonance phenomenon.* The previously mentioned change of sign in equation (3) when the resonance is crossed, which signifies a bending opposite to the original, corresponds to the well-known *phase jump* of the oscillation at the resonance, due to which the phase of the oscillation remains 180° behind the phase of the exciting force above the resonance. The self-centering of the shaft is explained by the general *inertial property of elastic systems*, which prevents them from following, to a perceptible degree, forces whose frequencies lie essentially above their natural frequency. In contrast to the following, it is emphasized that this natural frequency is independent of whether the turbine wheel rotates. Because of the placement of the wheel in the middle of the shaft, its rotational impulse is displaced only parallel to itself by the elastic oscillation, thus producing no gyroscopic effect and not at all influencing the character of the elastic oscillation. As a result, the eigenfrequency ω_k can be determined in advance by an experiment with the nonrotating shaft.

2. Numerical example.

A quantitative test of the theory developed here has the essential aim of proving the agreement between the critical angular velocity and the period of the eigenoscillation of the elastic shaft. This is accomplished by a simple experimental model*) with the following dimensions:

Shaft (silver steel):
thickness $d = 3$ mm,
length $2l = 40$ cm;
Disk (red brass), keyed at the middle of the shaft:
mass $M = 1$ kg.

The shaft is driven by hand, with an adjustable gear ratio (2, 3, or 4). The mass of the shaft is negligible compared with that of the disk.

The critical rotation number (that is, the number of rotations per minute in the state of maximum whirling) was observed with a metronome to be

$$n_k = 290.$$

*) The model belongs to the collection for technical mechanics of the Aachen Hochschule, and was constructed at the expense of the Aachen district of the Association of German Engineers.

The theoretical value of the same is, according to equation (4),

$$n_k = \frac{30}{\pi}\omega_k = \frac{30}{\pi}\sqrt{\frac{f}{M}}. \tag{9}$$

The magnitude of f, the measure of the elasticity of the shaft, can be determined directly by a bending experiment, as is done by F ö p p l. We are satisfied with the theoretical determination of f, and first use the above formula a). In this formula appear the quantities E and J.

The elastic modulus E of steel is approximately

$$E = 2 \cdot 10^6 \,\frac{\text{kg weight}}{\text{cm}^2} = 2 \cdot 10^6 \cdot 981 \,\frac{\text{kg mass}}{\text{cm sec}^2}.$$

The moment of inertia J of a circular cross section with diameter d and area F is calculated in a well-known manner as

$$J = \frac{F}{4}\left(\frac{d}{2}\right)^2 = \pi\frac{d^4}{64} = 4 \text{ mm}^4 = 4 \cdot 10^{-4} \text{ cm}^4.$$

There follows, according to a) with $l = 20$ cm,

$$f = 590 \,\frac{\text{kg mass}}{\text{sec}^2},$$

and thus, according to (9),

$$n_k = \frac{30}{\pi}\sqrt{590} = 230.$$

This value coincides in order of magnitude with the observed value of 290 rotations per minute, and deviates from it in the sense that we must expect from general considerations. In fact, formula a) gives only a lower bound for f and n, since the change of direction at the ends of the shaft is not completely free, but rather is restricted somewhat by the guidance of the bearings. If both bearings acted as completely fixed supports, which according to their light construction is certainly not the case, then, according to formula b), the value of f would be multiplied by four and the value of n would be doubled, and would considerably exceed the observed value.

A remarkable feature of the model is the significant work that must be applied to the hand crank if the shaft is to be maintained in rotation near one of the critical frequencies. Since the bearing friction (ball bearings) is very small and inertial resistance in continuous operation is generally not necessary to overcome, this work is applied solely for the overcoming of the internal friction that opposes the varying deflection of the shaft in the state of whirling. According to (3) or (8), the whirling is always greater, for equal rotational velocity, for greater e, and thus for greater original deviation of the shaft from straightness. For a somewhat perceptible curvature, it can occur that it is not at all possible to increase the angular velocity of the shaft beyond the critical velocity, even if, according to the specification of the chosen gear ratio, its rotation must lie above the critical value. The belt that connects the hand crank to the shaft then slips, in that its adhesion does not suffice to overcome the resistance to the motion of the whirling, and the shaft continues to rotate with a velocity somewhat below the critical velocity. This becomes especially striking if the hand crank is replaced by a motor of a few horsepower; even the motor cannot overcome the critical velocity, in that the belt fails. The shaft immediately runs smoothly, however, if one restricts the bending in the neighborhood of the critical state, and thus restricts the basis for the dissipation of work and the resistance to the motion, as was indicated on p. 888.

3. The occurrence of gyroscopic effects.

Gyroscopic effects did not enter the preceding discussion. They first occur if the turbine wheel is not only displaced by the whirling of the shaft, but is also rotated, or, dynamically expressed, if the rotational impulse of the wheel experiences a change of direction through the whirling. The conditions for this are especially favorable for the arrangement of Fig. 141, in which the turbine wheel is mounted at the end of a freely suspended shaft.

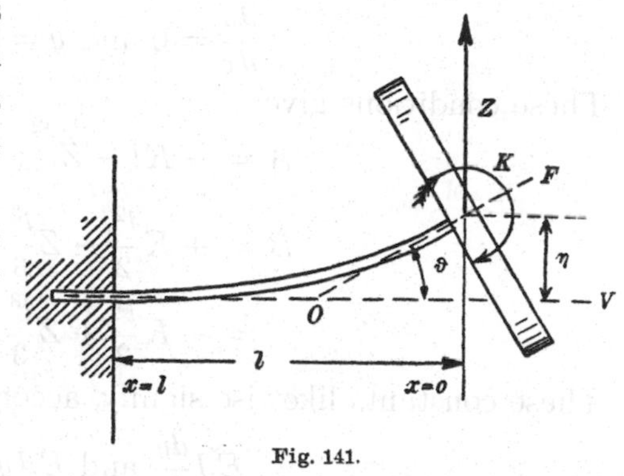

Fig. 141.

If we assume a uniform rotation of the shaft as the final state (after the damping of the free oscillations that are excited in the transient state), then this state consists, in the terminology of the theory of the top, of a uniform precession of the turbine wheel, in which its normal, the figure axis OF, is inclined by a certain angle ϑ with respect to the horizontal axis OV. The gyroscopic effect K of this precession stands perpendicular to OV and OF, and therefore acts about the normal to the plane of the drawing if OF lies directly in this plane. According to the rule of equi-orientational parallelism, it acts in such a sense that it tends to deflect the shaft back to its original position, just as the elastic resistance does; it is opposite to the centrifugal force Z.

We now ask for the equilibrium shape of our shaft when it is loaded at its end ($x = 0$) by the single-force Z and the force-pair K. This problem belongs to the theory of beam bending. The well-known fundamental equation of this theory is

$$EJ\frac{d^2y}{dx^2} = -\mathfrak{M}, \tag{10}$$

where y denotes the deflection from the original position (the x-axis OV) and

$$\mathfrak{M} = K - Zx \tag{11}$$

is the so-called bending moment at the cross-section x. From (10) and (11) there follow, by integration,

$$\left\{\begin{aligned} EJ\frac{dy}{dx} &= -Kx + Z\frac{x^2}{2} + A, \\ EJy &= -K\frac{x^2}{2} + Z\frac{x^3}{6} + Ax + B. \end{aligned}\right. \tag{12}$$

The constants of integration A and B are to be calculated from the boundary conditions

$$\frac{dy}{dx} = 0 \text{ and } y = 0 \text{ at } x = l.$$

These conditions give

$$\begin{aligned} A &= +Kl - Z\frac{l^2}{2}, \\ B &= +K\frac{l^2}{2} - Z\frac{l^3}{6} - Kl^2 + Z\frac{l^3}{2} \\ &= -K\frac{l^2}{2} + Z\frac{l^3}{3}. \end{aligned}$$

These constants likewise signify, according to (12), the magnitudes of

$$EJ\frac{dy}{dx} \text{ and } EJy \text{ at } x = 0.$$

If we write, according to Fig. 141, $-\vartheta$ for $\frac{dy}{dx}$ and η for y at the end of the shaft, then we have

$$\left\{\begin{aligned} EJ\vartheta &= -Kl + Z\frac{l^2}{2}, \\ EJ\eta &= -K\frac{l^2}{2} + Z\frac{l^3}{3}. \end{aligned}\right. \tag{13}$$

The quantities Z and K can be expressed in terms of η, ϑ, and the angular velocity ω. On the one hand,

$$Z = M\eta\omega^2,$$

and on the other hand, according to equation (III) of §1,*) if we replace $\sin\vartheta$ by ϑ and $\cos\vartheta$ by 1, identify in our case the velocity $d\psi/dt$ of the eigenimpulse N with the angular velocity ω, and calculate the eigenimpulse as $N = C\omega\cos\vartheta = C\omega$,

$$K = (C - A)\vartheta\omega^2.$$

Equations (13) thus give

$$(14)\qquad \begin{cases} \left(1 + \dfrac{(C-A)\omega^2 l}{EJ}\right)\vartheta - \dfrac{M\omega^2 l^2}{2EJ}\eta = 0, \\ + \dfrac{(C-A)\omega^2 l^2}{2EJ}\vartheta + \left(1 - \dfrac{M\omega^2 l^3}{3EJ}\right)\eta = 0, \end{cases}$$

for the determination of ϑ and η.

These equations yield, in general, $\vartheta = \eta = 0$ (that is, no bending) as the single possible form of motion of the shaft. We must have expected this, since we have tacitly assumed here that the shaft is perfectly straight—in contrast with No. 1, where we calculated from the beginning with a certain eccentricity e. Thus the centrifugal force and the gyroscopic effect vanish at the outset, and the straight state is theoretically maintained.

Nevertheless, equations (14) already have a significance for the possible bending of the shaft at the critical velocity. These equations, namely, are compatible with one another for arbitrary values of η and ϑ for the special case in which their determinant vanishes. This yields

$$(15)\qquad \left(1 + \frac{(C-A)\omega^2 l}{EJ}\right)\left(1 - \frac{M\omega^2 l^3}{3EJ}\right) + \frac{M\omega^2 l^3}{2EJ}\frac{(C-A)\omega^2 l}{2EJ} = 0$$

as the determining equation for ω^2. If we introduce the equidimensional ($\sec^2$) coefficients

$$(16)\qquad c = \frac{(C-A)l}{EJ}, \qquad m = \frac{Ml^3}{EJ},$$

then (15) becomes

$$cm\omega^4 - 12\left(c - \frac{1}{3}m\right)\omega^2 = 12.$$

The positive root of this equation is called ω_k^2. It is calculated as

$$(17)\qquad \omega_k^2 = \frac{1}{cm}\left(6c - 2m + \sqrt{12cm + (6c - 2m)^2}\right).$$

For this rotational velocity, the deviations η and ϑ are, as we said, un-

*) Here we must use the exact formula (III) instead of the approximation (II), since the rotation axis does not coincide with the figure axis of the turbine wheel, but rather with the "vertical" OV of Fig. 141. The rotation about the figure axis is thus first equal to $\omega\cos\vartheta$, but can be replaced, as is done in the text, by ω.

determined, and an arbitrarily strong whirling of the shaft can occur; for smaller and larger values of ω, in contrast, the shaft is exactly centered, as at the beginning. We thus obtain for η and ϑ the graphical representation of Fig. 142. At the point $\omega = \omega_k$ on the axis of the abscissa, the ordinate is undetermined, and is otherwise zero. One recognizes the meaning of this somewhat paradoxical result from other problems of mechanics (the buckling process, small oscillations of the pendulum). It signifies nothing other than a sudden increase and then decrease in bending for a crossing of $\omega = \omega_k$, a bending that in reality is naturally completely determined and depends on the original curvature of the shaft, but must remain undetermined for our starting point, since we have neglected the original curvature. The actual bending diagram will therefore have the form of the dotted line in the figure; for decreasing original curvature, it nuzzles always more into the rectangular corner that was calculated by us. The actual bending diagram therefore has the general form of the previous Fig. 140.

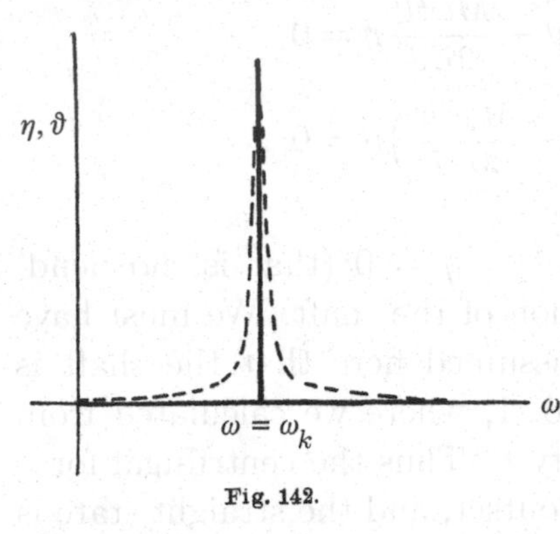

Fig. 142.

We can easily demonstrate this in detail if we supplement the present calculation by the introduction of the initial curvature, and thus adapt it to the calculation in No. 1.

We imagine, for example, that the initial curvature is produced by the influence of gravity on the overhanging turbine wheel, and therefore by a force Q at the point $x = 0$. We wish, however, to permit ourselves the imprecision of setting the moment of this force constantly equal to Qx for the entire rotation.

We therefore disregard, from now on, the special representation of gravity, and replace it, for our purpose, by a particularly convenient schematic force that co-rotates and acts, in principle, in the same way as gravity: to make, namely, the horizontal equilibrium position impossible by an initial curvature.

This simplifying assumption implies that we can directly carry over the preceding equations (10) to (13), where we now have only to insert $Z + Q$ instead of Z. Thus equations (13) are now

$$EJ\vartheta = -Kl + (Z+Q)\frac{l^2}{2}, \quad EJ\eta = -K\frac{l^2}{2} + (Z+Q)\frac{l^3}{3},$$

and one obtains, instead of (14),

$$\left(1+\frac{(C-A)\omega^2 l}{EJ}\right)\vartheta-\frac{M\omega^2 l^2}{2EJ}\eta=\frac{Ql^2}{2EJ},$$

$$+\frac{(C-A)\omega^2 l^2}{2EJ}\vartheta+\left(1-\frac{M\omega^2 l^3}{3EJ}\right)\eta=\frac{Ql^3}{3EJ}.$$

For each value of ω, there now follow entirely determined and generally finite values of ϑ and η. These values will be infinite only if the determinant of the left-hand side vanishes. This leads, however, exactly to the condition (15), and to the critical rotation number $\omega = \omega_k$. For decreasing Q (that is, for increasing straightness of the original form of the shaft), the just-named finite bending becomes zero, while the infinite bending for $\omega = \omega_k$ takes on the indeterminate value $0 \cdot \infty$, entirely in the sense of our Fig. 142.

It is of interest to us to establish the relative importance of the gyroscopic effect (the "composed centrifugal force," in the sense of C o r i o l i s) in comparison with that of the usual inertial effect (the simple centrifugal force) for the whirling of the Laval shaft. This is done in the simplest manner if we examine the degree to which the critical rotation number is increased because of the gyroscopic effect. (The gyroscopic effect indeed produces an increase and not a decrease of the critical rotation number, since, according to the rule of equi-orientational parallelism, it opposes the bending and therefore apparently increases the stiffness.)

We first disregard the gyroscopic effect, in that we set $C = 0$ and $A = 0$, and therefore take the radius of the turbine wheel to be very small. The resulting value of the critical velocity is called ω_{k0}. This results from (17) if we set $c = 0$. Since the form $0/0$ then appears in (17), we must make an expansion. For small c, with consideration of the positive sign of the root in (17),

$$\sqrt{12cm+(6c-2m)^2}=|6c-2m|\left(1+\frac{6cm}{(6c-2m)^2}+\cdots\right)$$

$$=2m-6c+\frac{6cm}{2m-6c}+\cdots,$$

and therefore, according to (17) and (16) for $c = 0$,

$$\omega_{k0}^2=\frac{3}{m}=\frac{3EJ}{Ml^3}.$$

The same value would also be taken directly from the second of the original equations (14), which for $C = 0$ coincides, mutatio mutandis, with equation (4).

In reality, A and C are not equal to zero, but rather are given, for a thin disk of radius r, by

$$A = \frac{M}{4}r^2, \quad C = \frac{M}{2}r^2;$$

thus, according to (16),

$$c = \frac{m}{4}\left(\frac{r}{l}\right)^2.$$

If we take, by way of example, the diameter of the disk equal to the length l of our shaft, there follow

$$c = \frac{m}{16}, \quad 2m - 6c = \frac{13}{8}m,$$

and, according to (17),

$$\omega_k^2 = \frac{2}{m}(-13 + \sqrt{217}) = \frac{3{,}4}{m}.$$

We therefore have

$$\omega_k^2 : \omega_{k0}^2 = 4 : 3{,}4; \quad \omega_k : \omega_{k0} = 1{,}09 : 1.$$

The critical rotation number is increased by 9% due to the gyroscopic effect. Its influence is therefore more of the secondary type.

Gyroscopic effects will occur not only for the freely suspended Laval shaft, but rather always when the whirling of the Laval shaft is bound with a change of direction of the rotating turbine wheel. This also occurs, for example, for a shaft with bearings at each end if the wheel is not directly in the center of the shaft, or if more wheels are placed on the same shaft. The calculation of the critical velocity then leads to a complicated system of equations, about which one can read everything worth knowing in the work of Stodola. New dynamic points of view do not appear; it is always a matter of equilibrium of centrifugal forces and gyroscopic moments on the elastic shaft. The numerical significance of the gyroscopic effect decreases as the turbine wheel comes closer to the center of the shaft, where the change of direction of the turbine wheel that is caused by the whirling becomes smaller. The case of the freely suspended shaft represents a maximum for the influence of the gyroscopic effect.

In conclusion, it may be mentioned that the whirling for these generalized arrangements can also be conceived as a resonance phenomenon; this conception, however, loses its practical significance, since the eigenoscillation period of the shaft is now influenced by the rotation, and thus the critical rotation number cannot be determined in advance by an experiment on the nonrotating system, just as the oscillational frequency of a top depends on its rotational frequency. For as soon as the direction of the rotating turbine wheel changes, the original bending oscillation is no longer a possible free motion of the system, since the gyroscopic effect acts as an external moment that changes the oscillation period. For each rotational velocity, one characteristic free oscillation period will be given; since this free period does not generally correspond to the rotation number, the shaft must experience a variable bending. The state of motion represented by equations (14) and (15) is also such a free oscillation; it is distinguished from others only by the correspondence of the rotational period and the oscillation number, so that the shaft can rotate as a whole. Thus the characteristic of the critical rotational velocity is here too the resonance between the rotation number and the free oscillation number.

§10. Miscellaneous applications.

Without striving in any way for completeness, we wish to summarize, in this concluding section, such applications of the theory of the top that we cannot discuss in depth, in part because the subjects do not appear sufficiently clear to us, and in part because they would lead us too far into technical details. We begin, as in §2, with gyroscopic effects for vehicles (single-rail trains), report on a few nautical problems, and conclude with the problem of ballistics, which, among all technical applications of the theory of the top, was conceived earliest and has been advanced least satisfactorily to the present day. An overview of the content is given in the following table.

A. Single-rail trains.
  1. The suspension railway.
  2. The monorail system.
  3. The system of Brennan and Scherl.

B. Nautical applications.
  4. The paddle steamer.
  5. The turbine steamer.
  6. The gyroscopic horizon.
  7. Remarks on aeronautics.

C. Ballistics.
  8. The dynamic-hydrodynamic problem.
  9. The general empirical facts of ballistics and the role of gyroscopic effects.

A. Single-rail trains.

The deliberations of §2 clearly show that the technical difficulties for high-speed trains—the conflicting and decisive economic difficulties of their introduction are not to be discussed here—depend primarily on the double-rail construction of our tracks. The two rails represent two kinematic guidance constraints, which, because of the necessary clearance for the wheels, are generally incompatible; the railcar must make, so to speak, a dynamic compromise between the two. Thus follow the alternating striking of the wheels against one or the other rail and the lurching of the carriage, the loading of one or the other rail by centrifugal and gyroscopic effects and the continuous deterioration of the track, and, in particular, the inconvenient necessity of a superelevation of the outer rail in a curve, which is calculated only for an average velocity and can never be fully implemented. All these evils vanish for a *single-rail train*, of which three types can be distinguished:

1. Stable arrangement: rail above the center of gravity of the carriage; *suspension railway*, patent by Lange.

2. Slightly stable arrangement: rail near the center of gravity of the carriage; *monorail system*, patent by Behr.

3. Unstable arrangement: rail beneath the center of gravity of the carriage; patent by Brennan and Scherl.

1. The suspension railway.

The first type has been implemented in Elberfeld–Barmen; it is regarded by many experts as the ideal of a technically perfect railway. The centrifugal forces and the gyroscopic effects of the driving and guiding wheels, which become rather small here because of the small radii of the wheels, can be balanced by lateral oscillations of the railcar without damage to the material. It may be remarked that the two effects do not add, as they do for the usual double-rail train, but rather subtract, which can be regarded as a desirable secondary effect of this arrangement. In fact, the moment of the centrifugal force about the rail, if the center of gravity lies *beneath* the rail, acts in the opposite sense as for the usual arrangement, and the moment of the gyroscopic effect acts in the same sense, and therefore (cf. §2) in the opposite sense to that of the moment of the centrifugal force.

We determine, for example, the swinging angle α of the carriage (the deviation of its midplane from the vertical) as it passes through a curve with radius of curvature R. The moment $\mathfrak{M}$ of gravity must be in equilibrium with the moment H of the centrifugal force and the gyroscopic effect K, so that

$$\mathfrak{M} = H + K. \tag{1}$$

If M denotes the mass of the railcar and h the height of the center of gravity beneath the rail, then

$$\mathfrak{M} = Mgh \sin \alpha.$$

If, further, v is the velocity in the curve, then

$$H = M\frac{v^2}{R}h \cos \alpha.$$

For the gyroscopic effect K, we must use equation (II) of §1, since the figure axis of the rotating wheels that produce the gyroscopic effect is inclined by the angle $\vartheta = \frac{\pi}{2} - \alpha$ with respect to the vertical, the axis of the added rotation (angular velocity v/R). We therefore have

$$K = -N\frac{v}{R} \cos \alpha;$$

the negative sign corresponds to the preceding remark on the opposite senses of the gyroscopic and centrifugal effects. The eigenimpulse N is, as in equation (2) of §2,

$$N = mvr,$$

where m denotes the mass of all the wheels reduced to their circumference, and r is their common radius; thus

$$K = -m\frac{v^2}{R}r \cos \alpha.$$

Our equilibrium condition (1) thus gives

$$\operatorname{tg} \alpha = \frac{v^2}{gR}\left(1 - \frac{m}{M}\frac{r}{R}\right). \tag{2}$$

For the trial runs in Barmen–Elberfeld, the velocity in the curves was increased until a swinging angle of 25° occurred; as the preceding formula shows, this angle would have been a little larger without the gyroscopic effect.[334]

2. The monorail system.*)

A monorail system with a speed of 150 km/hr was proposed as a high-speed train for the Manchester–Liverpool line; nothing has become known in recent years about the realization of this project. A test track with a speed of 135 km/hr was in operation for the occasion of the Brussels exhibition of 1897, and produced satisfactory results.[335]

The monorail system does not deserve the name "single-rail train" in the same measure as the suspension railway, since lateral guide rails, two on each side, are used beneath the load-bearing rail. The carriage forms a kind of saddle around the load-bearing rail, which it straddles with heavy extensions; it is supported on the bearing rail with a wheel in the vertical plane. In curves, or when the equilibrium is otherwise perturbed to one side or the other, the carriage lies against the guide rails on this side with two wheels in the horizontal plane. Such disturbances of equilibrium are to be expected, since the center of gravity of the entire carriage stands only a small distance from the running rail, and the carriage is therefore in a state of approximately indifferent static equilibrium.

Does this slight stability persist for full-speed travel, or does a more substantial stabilization of equilibrium occur, a stabilization that can obviously be produced only by a gyroscopic effect of the rotating wheels? According to information from the manufacturer, according to whom the lateral guide rails are required only exceptionally in curves, and according to the results of the trial runs, in which the repose of the carriage was praised, this must be assumed. It is indeed not inconceivable in itself that the rotating masses of the wheel and motor initially react to a lateral tipping of the carriage with a small rotation about the vertical, and that, by means of this rotation, a gyroscopic moment about the rail again uprights the carriage, as is described schematically by formula (IV) of §1. Since the gyroscopic effect increases as the square of the velocity of travel, an increase of stability with an increase of velocity (except with respect to the centrifugal force, which likewise increases as the square of the velocity) would be expected. However, the entirely essential condition of the required three degrees of freedom (cf. p. 767) must be considered. If the rotation of the wheels about the vertical is restricted by want of clearance or is significantly impeded by friction, the stabilizing effect also becomes impossible or is

*) Cf., for example, Centralblatt der Bauverwaltung 1899, p. 550.

strongly reduced. Whether a perceptible stabilization still occurs under these circumstances can be doubted. It is only certain that the stabilizing moment calculated according the ideal case of equation (IVe) in §1 will be much too large. It appears rather probable that stabilization can occur more easily for greater velocities than for smaller, since less clearance for lateral rotations of the wheels is required, and the friction that opposes such rotations not may increase with velocity in the same measure as the gyroscopic effect itself.

We find little satisfaction in this uncertainty, and, because of the small practical significance of the monorail system, would hardly have given it consideration here if it did not form a convenient bridge to the stabilization system for the unstable type of single-rail train. What is produced (or, as the case may be, not produced) by the rotating wheels only incidentally and under unfavorable circumstances in the present case, will be provided in the following case by an explicit gyroscopic device in an extensive measure and under deliberate conditions.

3. The system of Brennan and Scherl.

The question of the practical implementability of stabilizing a completely unstable single-rail carriage by a built-in rotor has often been discussed. But only most recently have experiments been conducted in the public view, and indeed simultaneously in November 1909 by L. Brennan[336] in England and A. Scherl in Berlin.[337] According to the reports, they have led to favorable results; concerning the detailed arrangement of the gyroscopic stabilizer design, however, nothing has become known, disregarding the earlier brief patent applications*) that touch upon the current form only in outline. We thus restrict ourselves here to a few generally held theoretical deliberations.**)

We have often discussed the real possibility of stabilizing an inherently unstable state by the gyroscopic effect, as, for example, in the discussion of the upright motion of the top. In principle, nothing other than this upright top motion, which naturally occurs here in a modified form, will be considered for the stabilization of the single-rail carriage.

*) Patent Brennan, D. R. P. 174402.

**) These deliberations are partly associated with an article by A. Föppl, "Zur Theorie des Kreiselwagens der Einscheinenbahn," Elektrotechn. Ztschr. 1910, 4.

We imagine, for example, a rotor in a single-rail carriage that is placed in a frame that may rotate about a transverse axis of the carriage, so that, as in the case of the ship stabilizer, the axis of the rotor can oscillate in the midplane of the carriage, and stands vertically in its mean position.*) The rotor then has, obviously, the particular degrees of freedom of a simple top in a Cardanic suspension: the frame of the rotor acts as the inner ring that may rotate about the transverse axis, and the carriage acts as the outer ring that may rotate about the track. The moments of inertia and the gravitational moments of the two noncyclic degrees of freedom are only different from one another here.

The first requirement for gyroscopic stabilization, that two noncyclic degrees of freedom must be present, is completely fulfilled here, in contrast to the monorail system. But the general theorem of Thomson cited on p. 771 requires still more: an even number of degrees of freedom is not sufficient for the cyclic stabilization of a unstable state, but rather an even number of unstable degrees of freedom is required.**) We can immediately conclude that if the rotor in the described arrangement is actually able to stabilize the carriage, it must, disregarding the gyroscopic effect, be itself in a unstable state, and its center of gravity must therefore lie above the suspension axis of the frame. For the ship stabilizer, in contrast, the center of gravity lay beneath the suspension axis ($h > 0$), since the ship is stable from the beginning ($H > 0$). We need not enter in more detail into the general proof of Thomson's theorem, since it is completely analogous to the following proof for the special case of the single-rail carriage.

In order to connect with the previously used notation for the ship stabilizer, let Q be the weight of the carriage, J its moment of inertia about the edge of the rail, and H the height of its center of gravity. Further, let q be the weight of the rotor and the frame, j its moment of inertia about the suspension axis, h the height of its center of gravity above the suspension axis, which we now assume as positive, and, finally, N the impulse of the rotor. The lateral rotation of the carriage is measured by the angle ψ, and the rotation of the frame of the rotor is measured

*) This is only one of various possible configurations. In the original implementation of Brennan (see below), the mean position of the rotor coincided with the transverse axis of the carriage. F ö p p l distinguishes three principal positions (see the previous footnote), but recommends the arrangement discussed in the text as the simplest.[338]

**) Whether stabilization is possible when one of the degrees of freedom is indifferent depends on special assumptions. We will return to this subject later.

by the angle ϑ. We assume, in the usual manner, that the investigation of small oscillations for ψ and ϑ is sufficient for the judgment of stability. We disregard frictional influences for the present.

The rotor will now be set into oscillation by lateral inclinations of the carriage. In addition to the gravitational moments, "gyroscopic effects" act on both degrees of freedom, as in the case of the ship stabilizer. The gyroscopic effect

$$K = N\frac{d\vartheta}{dt}$$

acts on the carriage, and the gyroscopic effect

$$k = -N\frac{d\psi}{dt}$$

acts on the frame of the rotor. We can directly carry over the equations of motion (5) of §4, and obtain, with the corresponding changes due to the original instability of the degrees of freedom ψ and ϑ,

$$(3)\qquad \begin{aligned} J\frac{d^2\psi}{dt^2} - QH\psi - N\frac{d\vartheta}{dt} &= 0, \\ j\frac{d^2\vartheta}{dt^2} - q\,h\,\vartheta + N\frac{d\psi}{dt} &= 0. \end{aligned}$$

One need only compare these equations with equations (5) of p. 367 in order to recognize the analogy with the upright motion of the top.

For a solution, we set

$$\psi = A \cdot e^{xt}, \quad \vartheta = a \cdot e^{xt},$$

and thus obtain

$$(3a)\qquad \begin{aligned} A(Jx^2 - QH) - aNx &= 0, \\ a(jx^2 - qh) + ANx &= 0, \end{aligned}$$

so that x satisfies the equation

$$(4)\qquad (Jx^2 - QH)(jx^2 - qh) + N^2x^2 = 0.$$

The coefficients in this equation are all positive, and thus x^2 can take on two negative values (and x four purely imaginary values), if

$$N^2 - QHj - qhJ > 0.$$

This requirement is supplemented by the actual stability condition, which is given by the discriminant equation

$$(N^2 - QHj - qhJ)^2 > 4QHqhJj.$$

In the case of the upright top, for which we can set $QH = qh = P$, $J = j = A$, the latter equation is transformed into the often used criterion

$$N^2 > 4AP$$

for the strong top.

The term $(-QH)(-qh)$ in (4) that does not contain x^2 is, as a product of two negative quantities, positive; it would be negative, however, if one of the two degrees of freedom were stable, and therefore if either H or h had a negative value. *Only if both degrees of freedom are stable or both degrees of freedom are unstable can the motion of the system be completely stable.* In just this manner, the last term of the corresponding equation for the general case of more than two noncyclic degrees of freedom will be the product of the generalized forces corresponding to the individual degrees of freedom; it will thus be positive if an even number of unstable degrees of freedom is present. Here lies the proof of Thomson's general theorem.*)

Of particular interest is the boundary case in which one of the two degrees of freedom (in our example, the motion of the frame of the rotor) is indifferent ($h = 0$). Such an arrangement is certainly to be excluded for the case of the single-rail carriage. For the characteristic equation (4) would always have, in that case, two, or, with the consideration of damping (see equation (6) below), one vanishing root. The corresponding integral $\vartheta = a + bt$, where a and b are constants, would state that the frame of the rotor could constantly rotate further in the same sense, and thus could move arbitrarily far from the vertical. With consideration of damping, the corresponding integral is $\vartheta = a$, which states that the frame of the rotor in this case, after an original oscillatory state is dissipated by friction, would remain standing in an arbitrary position, even, for example, horizontally, so that the axis of the rotor would be parallel to the edge of the rail. The rotor obviously then

*) In this generalized form the theorem holds, however, only for holonomic systems. For the proof carried out above for the case of two degrees of freedom assumes that all positions of the system have been expressed in terms of independent coordinates, and that the oscillation equations for these position coordinates are actually arranged in the Lagrangian form, conditions that are known not to be fulfilled in the nonholonomic case. The bicycle, for example, could also be stabilized if the center of gravity of the front wheel lay beneath the steering axis, and therefore if only one unstable degree of freedom, namely the inclination of the frame about the wheel track, were present. The smallest number of noncyclic degrees of freedom that permits of stabilization naturally remains equal to 2 in the nonholonomic case.

loses, however, any capability of stabilization. We made the same remark in the investigation of stabilization in §1, IV. There we assumed a moment Ψ that acts on the coordinate ψ, a moment that corresponds to the weight of the carriage for the single-rail system; the equilibrium with respect to the coordinate ϑ, in contrast, was completely indifferent. We found that the top can indeed stabilize the system (in the considered example, the outer ring of the Cardanic suspension) but that its capability of stabilization decreased with time and finally vanished completely.

The conditions are different for a gyrostat (cf. p. 771), or, still more simply, for an upright rolling disk, for which one degree of freedom, namely the rotation about the vertical, is likewise indifferent. That the characteristic equation there has two vanishing roots for sufficiently small friction implies that the disk can describe a precession with a constant inclination, a precession in which, for example, the center of gravity moves in a circle. Since the axis of the gravitational moment there is not fixed in space, but rather is constantly perpendicular to the figure axis, the figure axis can never coincide with the axis of this moment, and the rotation, which occurs approximately about the figure axis, retains an unweakened capability of stabilization. The same holds for the arrangement of the Thomson gyrostat. *The boundary case of indifferent arrangement is thus to be designated as stable in these cases.*

In the example of §1, the external moment is fixed in space, and the figure axis tends, according to the rule of equi-orientational parallelism, to approach the axis of the external moment. The stability thus decreases continuously, and is practically ensured only for a limited time. For the purpose of the torpedo, for example, this temporally bounded stability was sufficient; for the purpose and conditions of the single-rail train, in contrast, it would be completely insufficient. *The boundary case of indifferent arrangement of the frame of the rotor is thus certainly to be concluded as unstable for the single-rail train.*

We have thus far disregarded the influence of friction; in this case, the stability of the single-rail carriage for sufficiently large N was to be presupposed by analogy with the upright top. Essential difficulties, however, will naturally be caused by friction in the actual implementation. We consider friction by adding the usual terms in equations (3), without wishing to say more than that this corresponds to adding moments that always reverse their signs when the direction of the oscillation changes. Equations (3) then become

$$
(5) \qquad \begin{aligned} J\frac{d^2\psi}{dt^2} + W\frac{d\psi}{dt} - QH\psi - N\frac{d\vartheta}{dt} &= 0, \\ j\frac{d^2\vartheta}{dt^2} + w\frac{d\vartheta}{dt} - qh\vartheta + N\frac{d\psi}{dt} &= 0, \end{aligned}
$$

and there now follows, in place of equation (4),

$$
(6) \qquad \begin{aligned} &(Jx^2 + Wx - QH)(jx^2 + wx - qh) + N^2x^2 \\ &= Jjx^4 + (Jw + jW)x^3 + (N^2 - QHj - qhJ + Ww)x^2 \\ &-(QHw + qhW)x + QHqh = 0. \end{aligned}
$$

In order that stability be possible, all coefficients of this equation must again be positive. One sees from the coefficient of x that if both degrees of freedom are damped, a stable upright motion is not possible. *It is necessary that the oscillation of one degree of freedom be not damped, but rather accelerated.* It is inessential, in principle, which of the two degrees of freedom is chosen. In practice, however, only the possibility in which the frame of the rotor is accelerated may be considered. The idea of stabilizing the carriage by driving it farther in the direction of a lateral tipping is far too dangerous from a technical standpoint.

A device that continuously accelerates the oscillation of the rotor frame is in fact present in the patent of Brennan. The quantity w is thus to be imagined there as negative. The coefficient of x in equation (6) can then become positive, but, on the other hand, the coefficient of x^3 would become negative if W were not at the same time sufficiently large and positive. *It therefore does not suffice to accelerate one degree of freedom; the other must be simultaneously damped in a corresponding measure.* If we set $w = -w_1$, then the conditions to be fulfilled are

$$
(7) \qquad \begin{aligned} qhW - QHw_1 &\geqq 0, \\ jW - Jw_1 &\geqq 0. \end{aligned}
$$

The two conditions are compatible if

$$
\frac{qh}{j} \leqq \frac{QH}{J}
$$

is chosen.*) And indeed, according to the second inequality of (7), W must have a much greater value than w_1, because of the very large gravitational and inertial moments of the carriage with respect to those of the rotor. According to the first inequality of (7), however, a relatively

*) This implies that in a hypothetical reversal of the direction of gravity, for which the oscillations of both the carriage and the rotor would be stable, the frequency of the rotor frame must be smaller than that of the carriage.

small acceleration of the oscillation of the rotor frame suffices; for what concerns the damping of the carriage oscillation, this will always be relatively large in itself, or can be artificially increased. One best sees that stability can in fact be achieved by such an arrangement if the equal signs in equation (7) are assumed; equation (6) then has the form of equation (4), which implies, for sufficiently large impulse, stable undamped oscillations. One now easily understands that for the case of the inequalities in (7), stability not only persists, but rather, as it is to be required, is also ensured to a higher degree, since initially introduced oscillations will be damped.

The inequalities (7) also show why one will not consider accelerating the carriage and damping the frame of the rotor. Were W negative and w positive, and therefore w_1 negative, the conditions (7) would be fulfilled under the assumption $qh/j \gtreqless QH/J$, but, according to the first inequality, the absolute value of W must then, because the gravitational moment QH of the carriage far exceeds the gravitational moment qh of the rotor, be much larger than that of w, and therefore either the carriage must be very strongly accelerated or the frame of the rotor must be very lightly damped, two possibilities that will evidently lead to difficulties.

The requirement that one degree of freedom must be accelerated in order for stabilization to be possible has been derived here analytically from the equations. Another point of view makes this somewhat astonishing requirement more understandable: accelerating the frame of the rotor strengthens the moment carried over from the rotor to the carriage, the moment that is able to upright the carriage. The uprighting moment on the carriage is indeed given only by the term $Nd\vartheta/dt$ in equations (5). We found that it is never possible to damp initially introduced oscillations by only strengthening the eigenimpulse, and, with consideration of the inevitably present (even if small) friction, that stabilization cannot be achieved. Because of the factor $d\vartheta/dt$ in the gyroscopic moment, it is natural to enlarge, so to speak, the lever arm of the gyroscopic effect, and thus to to accelerate the degree of freedom ϑ.

The energy thus introduced into the system finds a twofold application: first, energy is required for the elevation of the center of gravity into the upright position after an initial lateral inclination of the carriage. Second, the potential energy of the carriage is transformed into kinetic energy in any oscillation, and as such is partly dissipated by friction; it must be replaced in order that the carriage be able to return to the upright position. On the other hand, however, an actual damping of one degree of freedom is required to dissipate the kinetic energy that is imparted to the carriage by an impact. Thus follows the necessity of damping the degree of freedom ψ.

The motion of the gyroscopically stabilized carriage may be described intuitively by comparison with the oscillation of the upright top; that is, by comparing it with pseudoregular precession when the precession circle is small (cf. p. 342).*) In both degrees of freedom, the complete motion is a superposition of two oscillations, where the oscillation of the rotor frame is displaced by approximately half an oscillation period with respect to that of the carriage, as follows from equations (5) for small damping. As in the case of the upright top, we denote the slower oscillation as a precession and the faster as a nutation. In our case, both oscillation curves are, if we disregard damping, ellipses whose midpoints lie on the vertical; under the particular conditions of the upright top they become circles. For sufficiently large impulse, we calculate the small frequency of the first oscillation by neglecting the powers x^4 and x^3 in (6), and calculate the large frequency of the second by neglecting the power x and the constant term. According to the sign convention for w and W and the inequalities (7), the positive term $-QHw = +QHw_1$ is decisive for the damping of the first oscillation and the positive term jW is decisive for the second oscillation, while the term $Jw = -Jw_1$ would remain and make the damping factor of this rapid oscillation negative if W were not present.

We thus see that the measure of accelerating one degree of freedom suffices only to damp the precessional oscillation, but that the nutational oscillation constantly grows in time and makes the system unstable if it is not artificially damped or sufficiently reduced of itself by rail friction.

In the patent of Brennan, several devices are provided for the acceleration of the precession that we have considered schematically by

*) This comparison is indeed made difficult by the fact that the meaning of our current coordinates ψ, ϑ is entirely different from the meaning of the previous ψ, ϑ. The angle ψ, as defined on p. 905, is now measured about the horizontal and spatially fixed rail, and the angle ϑ is measured about the approximately horizontal transverse axis that is fixed in the carriage. In contrast, ψ was previously measured about the spatially fixed vertical, and ϑ about the exactly horizontal line of nodes.

the term $-w_1\frac{d\vartheta}{dt}$ in equations (5). The phenomenon on which these devices are based is known from the simplest experiment: a top placed on its tip, which thus describes a slow precession, begins to roll rapidly if its rotating disk touches the support surface. In a similar manner, the rotor of the single-rail carriage can roll on lateral surfaces that are applied against its circumference. The device is made so that when the direction of the inclination of the carriage changes, the side on which the surface is applied also changes; the direction of the rolling thus changes with the sign of ψ. Now according to equation (3) (in which damping is not considered), the inclination ψ is in phase with the direction of the oscillation $d\vartheta/dt$. Thus the rolling, with appropriate determination of the sign, will always take on the sense of the oscillation of the rotor frame, and this oscillation can be accelerated. The energy of the acceleration is taken directly from the rotor, and indirectly from the driving motor. In practice, it is naturally advantageous to use not the rotor itself, but rather a roller appropriately attached to it, as is done in the design of Brennan. One may also easily design other devices for accelerating the oscillation of the frame; for example, an arrangement similar to the Anschütz damping device (cf. p. 861).

The upright arrangement of the rotor discussed here is not adopted in the original patent of Brennan; the figure axis there oscillates not in the vertical plane, but rather in the horizontal plane, so that it must lie transverse to the carriage in its mean position. Its equilibrium position is then indifferent, and it must, according to general principles, first be made artificially unstable, perhaps through spring- or weight-loading. In this case, furthermore, a change of position of the impulse must obviously occur in a turn, and thus moments that can cause strong oscillations are transmitted to the carriage, while a curve does not directly influence the rotor in the vertical position. In the horizontal arrangement, two rotors with opposite senses of rotation and coupled frames are used, so that the damaging gyroscopic effects are canceled, as Mr. Skutsch has proposed for the ship stabilizer. The two arrangements are otherwise not different from one another in principle.[339]

The deliberations thus far carry over in an analogous manner to the case in which the carriage is influenced not only by gravity, but also by the centrifugal force in a curve. Its equilibrium position is then naturally determined, as for the two-rail carriage, by the condition of equilibrium between the gravitational moment and the moment of the centrifugal force.

This equilibrium position is inclined inward; the stabilized carriage places itself, to the extent that a stationary state is approached in a long curve, at the correct inclination, just as a two-rail carriage is inclined artificially by the superelevation of the outer rail. The attainment of a stationary state naturally depends on stabilization by the rotor, by means of which the now inclined equilibrium position is ensured in exactly the same manner as the upright position on a straight track. The theoretical treatment of the positioning would differ from the preceding only by the referral of the equations to the new equilibrium position instead of the mean position $\psi = 0$. The positioning at the correct inclination will indeed begin with an inclination toward the outside of the curve, which is caused by the primary effect of the centrifugal force. This initial inclination, however, is immediately reversed by the gyroscopic effect, and, by means of damped oscillations, is transformed into a final inclination toward the inside. If the transition to the greatest curvature of the rail is sufficiently long, the state of the carriage can be conceived as a continuous, slowly changing state of equilibrium.

This phenomenon of immediate self-alignment is, in fact, confirmed by the eyewitnesses of the trial runs in Berlin; they likewise confirm that in the case of a one-sided loading, which indeed acts similarly to a centrifugal force, the carriage inclines immediately to the opposite side and comes to the equilibrium position in which the now displaced center of gravity of the entire system lies exactly above the rail. The primary effect of the loading is naturally an imperceptibly small yielding, which, however, is immediately followed by the uprighting gyroscopic effects that stabilize the new equilibrium position.

Strong perturbations or impulses of short duration that are transmitted to the carriage in travel may be damaging for operational safety. Let, for example, L be the magnitude of such an impulsive moment about the rail, which in the mean will be known to the designer. The deflection with which the carriage will react to such a disturbance can be obtained from a simple deliberation. If the forward motion is disregarded, the eigenimpulse of the carriage is the vertically (for the first arrangement) directed quantity N. If we add the perturbing impulse L

vectorially, the axis of the resulting impulse will be inclined in the midplane in the direction of the angle ϑ by the amount

$$\operatorname{tg}\vartheta_L = \frac{L}{N}. \tag{8}$$

We could treat of the resulting motion in detail, exactly as we performed the calculation for the impulsively generated motion of a ship (p. 802). We can also, however, give a simple geometric overview of the result. We previously (p. 910) described the motion of the rotor, namely, as the superposition of a precession and a nutation. According to the previous detailed discussion of pseudoregular precession, the half-opening angle of the nutation cone (that is, the nutation amplitude) is given directly as the angle between the initial position of the figure axis, which stood vertically, and the impulse axis; it is therefore equal to the angle ϑ_L. This angle initially signifies, since the nutation cone is an ellipse, only the major axis in the ϑ-coordinate. The mean axis of the nutation cone (namely, the impulse axis) gives, at the same time, the amplitude of the precession, which therefore, measured by the ϑ-coordinate, likewise amounts to ϑ_L. The full deviation of the rotor frame from the vertical, which at its maximum is the sum of the precessional amplitude and the nutational amplitude, will therefore amount at most to $2\vartheta_L$. As in the case of the ship stabilizer, the total deflection of the rotor frame must be limited to approximately 45°, and therefore $\operatorname{tg} 2\vartheta_L$ may not exceed the value 1. For safety, one must therefore demand that

$$N > 2L,$$

where we have used ϑ_L instead of $\operatorname{tg}\vartheta_L$ in (8).

In order to estimate the resulting deflection of the carriage, we restrict ourselves to the case of equal "frequency" of the two degrees of freedom, and thus to the assumption

$$\frac{QH}{J} = \frac{qh}{j}.$$

The ratio of the rotor and carriage oscillations then follows immediately by division of equations (3a) as

$$\left(\frac{A}{a}\right)^2 = -\frac{j}{J},$$

and thus (8) gives the amplitude of the precessional oscillation of the carriage, under the assumption that it is small, as

$$\psi_L = \frac{L}{N}\sqrt{\frac{j}{J}}.$$

The total amplitude is at most $2\psi_L$.

This deliberation may be considered as a rough approximation for the required design quantities if the greatest perturbing impulse L and the greatest permissible deflection $2\psi_L$ are known.

This consideration applies, moreover, to both the first and second arrangements. What forms and dimensions will prove favorable in particular cases, only the future of single-rail trains will tell.

B. Nautical applications.

We summarize, in the following, a few problems of ship travel that are associated with the theory of the top. The effects discussed under No. 4 are hardly new; they concern only relations already known to us from the railway, carried over to the paddle steamer. The loads on the bearings in a turbine ship considered in No. 5 may have greater interest. The gyroscopic horizon in No. 6 is a device that pursues similar goals as the gyroscopic compass in §8, but with essentially more modest means and correspondingly more limited results.

4. The paddle steamer.

When the rudder of a ship is deployed, for example, to the starboard (that is, to the right of the direction of travel), the ship describes a curve whose center of curvature lies to the right. In this curve, the long direction of the ship does not coincide exactly with the direction of the trajectory of the center of gravity, but rather is rotated by the so-called drift angle toward the side of the center of curvature.[340] The magnitude of this angle amounts in the mean, for different types of ships and different velocities, to $\delta = 10°$.

The forces that act on the ship are the following.

1. The water resistance that acts along the length of the ship for travel in a straight line, but is inclined to the long axis in a turn.

2. The rudder pressure that acts perpendicularly against the surface of the rudder.

3. The paddle wheel pressure that arises from the driving force of the engine and acts along the length of the ship.

4. The gyroscopic effect of the paddle wheels. The axis of this moment stands perpendicular to the axis of the wheels and the vertical, and therefore falls along the long axis of the ship; it acts, as seen in the direction of travel, in the counterclockwise sense. The magnitude of the gyroscopic effect, in the often used notation, is $Nd\psi/dt = mv^2r/R$, where R denotes the radius of curvature of the trajectory of the center

of gravity and m denotes the reduced mass of the two wheels and the engine parts that rotate with them, and indeed reduced to the distance r from the rotation axis at which the rotational velocity of the wheel is directly equal to the velocity of travel v.

In the process of steering, the ship begins to turn about the vertical in the direction of the drift angle δ at the moment when the rudder pressure is first applied. Since the direction of travel no longer coincides with the long direction of the ship, the water resistance now produces a moment about the vertical that tends to turn the ship, because of its stable form of construction, back to the direction of travel. The definitive value of the angle δ is attained when the moment of the water resistance is equal and opposite to the moment of the rudder pressure. At the same time, the water resistance and the paddle wheel pressure no longer completely cancel each other, but there remains a component perpendicular to the direction of travel, which has as a consequence the lateral acceleration. Its magnitude must be equal and its direction opposite to the centrifugal force that corresponds to the curvature of the line of travel; the magnitude is thus

$$W = \frac{Mv^2}{R},$$

where M denotes the total mass of the ship. The point of application lies approximately in the middle of the draft and at the midlength of the ship. To a smaller degree, moreover, there is a contribution from the rudder pressure that we can imagine, after the separation of its turning-moment about the vertical, to be applied at approximately the same point; it is directed, however, away from the center of curvature. We imagine the component of the rudder pressure perpendicular to the direction of travel as calculated together with W. As is well known, the ship leans somewhat to the side when it is steered; it *heels*, as the technical expression goes. Most types of ships heel outward from the center of curvature. For the heeling of the ship, the transverse component $W \cos \delta$ of our force W comes into consideration.[341]

The center of gravity of the ship is generally higher than the point of application of the named force. If the difference in height is l, the latter has a moment $lW \cos \delta$ about the center of gravity that inclines the ship to the outside. The resulting heeling angle α in a stationary state is now determined by the condition that this moment, strengthened by the moment of the gyroscopic effect, maintains equilibrium with the uprighting gravitational moment $Mgh\alpha$, where h denotes the metacentric height. Written explicitly, this equation is

$$Mv^2\frac{l}{R}\cos\delta + mv^2\frac{r}{R} = Mgh\alpha.$$

The heeling angle α is thus determined as

$$\alpha = \frac{v^2}{gh}\frac{l}{R}\left(\cos\delta + \frac{m}{M}\frac{r}{l}\right). \tag{9}$$

This equation is analogous in form to equation (2) of this section. The opposite sign of the second term in the parentheses, which represents the relative influence of the gyroscopic effect, depends on the gyroscopic effect having the same sign as the moment of the centrifugal force for the approximately upright position of the ship, while the corresponding moment was of the opposite sense for the downward-hanging carriage of the suspension railway. Here, as there, it is apparent that the influence of the gyroscopic effect is small, particularly, according to equation (9), since the reduced mass m of the wheels is very small in comparison with the total mass M of the ship.

On the other hand, a favorable influence of the gyroscopic effect on the stability of the paddle steamer with respect to rolling motion has often been conjectured. This influence would be entirely analogous to the likewise hypothetical influence on stability for Behr's single-rail carriage in No. 2, and its theoretical confirmation would confront the same difficulty as there: the strong impediment to the turning of the ship or the carriage about the vertical due to water resistance or rail friction. Nevertheless, this alleged autonomous stabilization of the paddle steamer can claim a certain historical interest, as it was cited by Schlick*) and appears to have been the starting point for his design of the ship stabilizer.

5. The turbine steamer.

The most recent type of ocean liner, the turbine steamer, creates gyroscopic effects of much larger magnitude than those of the paddle steamer. The large angular velocity of the steam turbine, in comparison with the moderate velocity of the paddle wheel, particularly strengthens the gyroscopic effect. Since the shaft of the turbine is placed along the length of the ship so that it can be coupled directly to the ship's propeller, the axis of the gyroscopic effect is turned by 90° with respect to its position for the paddle steamer. A rolling motion of

*) Cf. his previously (p. 808) cited presentation to the Schiffbautechnische Gesellschaft, p. 121 ff.

the ship (a rotation about its long axis) produces no gyroscopic effect. A pitching motion (a rotation about the transverse axis) produces a gyroscopic effect about the vertical. In a maneuver of the ship (a rotation about the vertical), a gyroscopic effect occurs about the transverse axis.

At the beginning of turbine ship construction, it was often feared that a turbine ship would be more difficult to steer than a screw steamer. The development of the matter has not justified this fear. We wish to examine the subject theoretically. The immediate consequence of steering (turning angle ψ), which may be initiated by a turning moment Ψ about the vertical, is a gyroscopic effect $K = N\frac{d\psi}{dt}$ about the transverse axis, where N now signifies the impulse of the turbine (the angular velocity times the moment of inertia of the entire system of rotating masses). This gyroscopic effect acts to cause a pitching of the ship (rotation angle ϑ). The pitching produces a gyroscopic effect K' about the vertical of magnitude $N\frac{d\vartheta}{dt}$, which opposes the moment Ψ of the rudder pressure. The angular velocity $d\vartheta/dt$ of the pitching, however, is small for two reasons: the enormous moment of inertia of the ship about the transverse axis, which here takes on its maximum value, and the water resistance that restricts the pitching. Only to the extent that the ship is freely mobile about the transverse axis can the countermoment against the steering (also a kind of stabilization against external moments) occur. When the pitching is restricted, we again have a top with only two degrees of freedom and no intrinsic capability of resistance. This entire question, moreover, comes into consideration only for an asymmetric design with only one turbine shaft.

Only the loading on the bearings due to the gyroscopic effects is seriously considered today;*) this has been cited by experts as a possible cause of actual accidents in turbine torpedo boats.[342] We assume a perceptibly rigid shaft in the body of the turbine (corresponding to the Parsons or Curtis turbine primarily used in ship construction), not a flexible shaft (cf. the preceding section on the Laval turbine), for which, in essence, a bending of the shaft occurs instead of the full deflection of the turning impulse, which generally corresponds to a smaller load on the bearings. Moreover, a small part of the gyroscopic effect

*) A. Stodola, Die Dampfturbinen, Nr. 104.

may be eliminated by elastic bending even for a perceptibly rigid shaft, so that the entire load to be calculated here does not act on the bearing.

The entire gyroscopic effect is a force-pair $N\frac{d\psi}{dt}$, whose two single-forces act on the two bearings of the turbine shaft in opposite directions, and indeed in the vertical or horizontal direction, according to whether the gyroscopic effect is due to steering or pitching.

We consider at once a numerical example.*) The turbine makes 250 revolutions per minute, corresponding to an angular velocity of $\omega = 2\pi \cdot 250/60$. The mean diameter of the rotors is 2,86 m. The weight of the rotating parts reduced to this diameter amounts to approximately 18 000 kg, and the distance between the bearings is 5,55 m.

As the moment of inertia, one obtains

$$\begin{aligned} J &= 18000 \cdot 1{,}43^2 \text{ kg (mass) m}^2 \\ &= 1800 \cdot 1{,}43^2 \text{ kg (weight) m sec}^2 \\ &= 3600 \text{ kg (weight) m sec}^2, \end{aligned}$$

and as the eigenimpulse

$$\begin{aligned} N = J\omega &= 2\pi \cdot \frac{250}{60} \cdot 3600 \\ &= 2\pi \cdot 15000 \text{ kg (weight) m sec.} \end{aligned}$$

For an angular velocity of 10° per second $= \frac{\pi}{18}$ sec^{-1} for the steering or pitching motion, the gyroscopic effect is

$$K = N\frac{\pi}{18} = \frac{50000}{3} \text{ kg m.}$$

This moment is distributed on the two bearings with a lever arm of 5,55 m. The load on each bearing is therefore

$$P = \frac{50000}{3 \cdot 5{,}55} = 3000 \text{ kg.}$$

This is only a small fraction of the constant loading due to the weight of the turbine (greater than half of 18000 kg). Our calculation therefore shows that the danger of such an additional bearing load due to the gyroscopic effect appears not to be of immediate significance.

*) This example corresponds to a large ship turbine, the 4000 hp Curtis system of the steamer “Creole”; cf. Engineering, 1906, p. 696.[343]

6. The gyroscopic horizon.

The problem of determining geographical location, especially the measurement of the elevation angle of a star, assumes the establishment of the horizontal plane. While this plane can be determined on land by a spirit level or a surface of quicksilver, there remains only the observation of the natural horizon on an oscillating ship. This becomes impossible, however, in fog or on a stormy sea. There follows, therefore, the problem of constructing a constantly observable artificial horizon.*)

Captain Fleuriais**) uses, for this purpose, a top with the following manner of construction: copper ring (weight 175 gr) with a bell-shaped continuation downward; short figure axis with a conical end that rests on a seat; center of gravity ca. 1 mm beneath the support point; pneumatic actuation (electric actuation was not yet available then); rotation number per second approximately 80 at the beginning and 50 at the end of the observation; actuation not continuous, but rather only at the beginning of the observation. The diminishment of the rotation number is due essentially to air resistance, so that it was also attempted to let the apparatus run in an evacuated chamber. On the upper plane surface of the copper ring, two plano-convex lenses L, L' are attached to the ends of a diameter, arranged so that the one forms the plane surface of the other in the observation telescope. Each lens carries on its plane surface a streak that runs perpendicularly to the figure axis of the rotor, and is therefore horizontal when the figure axis is upright. When the top rotates, the observer receives two impressions in the telescope during each rotation, alternatively from the streaks of the lenses L and L', and outlined by the lenses L' and L, which for a large rotation number merge into one image. This image is a horizontal streak if the figure axis is vertical, and directly represents the trace of the horizontal plane. In reality, the figure axis on the oscillating ship is not vertical, but rather slowly describes a precession cone under the influence of gravity; its period, for the dimensions of the apparatus, is

*) Experiments with this goal were made as early as a century and a half ago; cf. Serson, Philosoph. Transactions, London 1752, and The Gentleman's Magazine 1754.[344]

**) Fleuriais, Bulletin astron. 3, 1886, p. 579; De Jonquières, Comptes Rendus, t. 104, Paris 1887; Baule, Revue maritime, 1890.

approximately $1^1/_2$ minutes. The short nutations that are superposed on the regular precession do not come into consideration for the observation in the telescope; the accelerating forces due to the oscillations of the ship, which are of a much shorter period than the precessional period of the top, will also, for the most part, average out. The impression of the observer is thus essentially the same as for an exact precession; that is, a horizontal streak appears in the field of view if the figure axis is directly in the vertical plane through the axis of the telescope. But an image of the streak also results for all intermediate positions of the figure axis (or, for each rotation, two images that merge with each other); this image, however, is now inclined, corresponding to the instantaneous position of the figure axis with respect to the horizontal. The passage from the upper to the lower horizontal position of the streak is therefore influenced by the successively changing and inclined positions of the streak. The observer must essentially note the two horizontal positions; he finds the trace of the horizontal plane passing through the viewpoint by constructing their midline. This bisecting line is the desired *gyroscopic horizon.*

For an actual observation, it is better to take the mean of not two, but rather three successive horizontal positions of the streak, since the figure axis slowly uprights due to friction in the seat and air resistance. A sextant is attached to the telescope for the determination of the elevation angle. Its mirrors are placed so that the image of the star in the telescope coincides exactly with the established position of the gyroscopic horizon; the elevation angle can then be directly read as twice the rotation angle of one of the mirrors.[345]

The only objection that can be raised against this ingenious construction is the difficulty in handling. An expert observer obtains a precision of a few minutes; months of practice are required for this. We ourselves had the occasion to make a measurement with the gyroscopic horizon at the Kiel observatory, and can confirm the difficulty of its use for the inexperienced. Many realizations of the device have been used in the French navy; the Russian fleet in the Russo-Japanese war was also so equipped.

7. Remarks on aeronautics.

The role of gyroscopic effects in air travel has naturally been much discussed in recent times.*) We think first of the direct effects of the propeller on a motorized aircraft. These effects are the same as those on the turbine steamer, and, disregarding the direction, on the paddle steamer and the monorail, but the effects for the relatively lightly built aircraft are much more important, since the rotating masses make up a much greater fraction of the total mass of the system. A lateral steering of the aircraft will thus be accompanied by a tipping, and an ascent or descent of the long axis by a lateral deflection. This phenomenon, as we have heard and is not to be otherwise expected, is actually observed.

On the one hand, such effects will be regarded as undesirable, since they detract from the maneuverability of the aircraft (and here, indeed, in a completely different order of magnitude than for the turbine steamer in No. 5); on the other hand, they can be discussed as a desirable contribution to the stability of the aircraft. If one pursues, namely, the coupling of these perpendicularly and laterally deviating moments as in §1 IV, one is easily convinced that here too the direction of the propeller axis will be strongly stabilized, for a sufficiently rapid rotation, against any type of overturning moment with a perpendicular axis. Whether the favorable or unfavorable gyroscopic effects of the propeller predominate is to be decided from case to case, according to the manner of construction of the device. The Wright brothers appeared to fear that the unfavorable effects would predominate, in that they used a pair of counterrotating propellers.

On the other hand, the possibility of a special gyroscopic stabilization of aircraft has already been conceived. Since aviation has had to struggle with essential difficulties in stability, one cannot deny an actual significance to such a construction. One can distinguish here, as for the torpedo, a direct and an indirect stabilization. We refer, in particular,

*) Cf., for example, the articles of L. Prandtl, "Einige für die Flugtechnik wichtige Beziehungen aus der Mechanik," in the Ztschr. f. Flugtechnik u. Motorluftschiffahrt, Jahrg. I, Heft 1–7, 1910.[346]

to a note of Carpentier*) that describes a model built by Regnard. Here the indirect principle is applied (cf. the Whitehead torpedo, §3): a top that does not take part in the oscillations of the aircraft activates an electrical contact that causes the deployment of a counter-rudder. It is not known to us whether the direct principle of the fixed or oscillating rotor (as in the Schlick stabilizer) has already found experimental application.

C. Ballistics.

Among all problems of terrestrial mechanics, the problem of ballistics was perhaps the first to be treated by the methods of dynamics.**) The ballistic curve was sought by d'Alembert and Euler. The ballistic rotation problem was conceived theoretically by Poisson and experimentally by Magnus. Even today we involuntarily think of the projectile in air, at whose center of mass gravity acts, as the "freely moving rigid body of dynamics." We illustrated on p. 533 how far we remove ourselves from reality if we do not simultaneously consider air resistance. In the theoretical treatment of this decisive quantity for external ballistics, the problem properly lies not in the domain of rigid body dynamics, but rather in that of hydrodynamics of compressible fluids, or, more correctly, in a particular coupling of the two domains. We wish to consider this problem in No. 8. Since an actual solution of the ideal ballistic problem can hardly be expected in the near future because of its extraordinary mathematical difficulty, one is led to approximate and empirical considerations for the various influences of the surrounding air medium. As these influences are combined with the dynamics of rigid bodies, special emphasis is placed in No. 9 on the alleged role of gyroscopic effects.

According to what has been said, it is clear that the difficulty of the ballistic problem lies rather on the hydrodynamic than the dynamic side. One may perhaps hope that the commencement of air travel will give new life to hydrodynamic problems, and also be of benefit to ballistics.

*) Comptes Rendus, t. 150, p. 829, March 1910.[347]

**) Extensive literature references are given by C. Cranz, Encykl. d. math. Wiss. Bd. IV, Art. 18, or Lehrbuch der Ballistik, Leipzig 1910, Teil 1, external ballistics.

8. The dynamic-hydrodynamic problem.

It is self-evident that a body in a fluid, strictly speaking, has not six but rather infinitely many degrees of freedom. The exact determination of its motion (its six coordinates) is possible only in conjunction with that of the fluid (its infinitely many position parameters). In addition, the two components of the motion, the translation of the center of gravity and the rotation about the center of gravity, which in the absence of the external medium (that is, with the neglect of its inertia, etc.) can be treated independently, are inseparably bound to one another by the simultaneous motion of the surrounding fluid.

The exact formulaic dependence between the motion of the fluid and the motion of the body is well known to the mathematician for the case in which the surrounding fluid is imagined as incompressible and frictionless, and its motion as irrotational. The incompressible medium is positively bound, so to speak, to the moving body; since this medium transmits all disturbances with an infinitely large speed of propagation, the instantaneous state of motion of the body has an immediate effect in the entire infinite fluid; the fluid motion depends only on the instantaneous value of the velocity of the body, and has no memory of earlier states. In this case, therefore, the *vis viva* of the entire system, the body and the fluid, can be represented as a function of the instantaneous values of the six velocity coordinates of the body, so that the equations of motion of the system can be obtained according to Lagrange's method in the form of six ordinary differential equations of the second order. Hydrodynamics acts here only in the determination of the *vis viva* of the system, which now, for example, can no longer be composed from the mere superposition of a translational and a rotational component; the formation and character of the differential equations, in contrast, correspond to ordinary dynamics.

In reality even water, the prototype of an incompressible fluid, has a finite speed of sound, and is therefore compressible. One can, however, treat any gaseous fluid as incompressible as long as the greatest velocities present are small compared with the speed of sound. The influence of previous states of motion then propagates very rapidly and dissipates at infinity as new disturbances are produced by changes in the state of motion. Everything thus depends perceptibly on the instantaneous state of motion, as for ideal incompressible behavior.

While the simplifying assumption of incompressibility may be generally permissible for the theory of aircraft, the criterion for quasi-incompressible behavior is unfortunately not at all fulfilled for ballistics. The initial velocity of an artillery or infantry projectile is 465 or 885 m/sec, respectively, and is therefore "supersonic." For long-range artillery fire, the projectile passes through the critical value of velocity in the course of the shot and lands, for a firing range of 4000 m, with the "subsonic" velocity of 257 m/sec; the velocity of a bullet from an infantry rifle decreases to 166 m/sec for a firing range of 2000 m. It is known, however, that the flow field is fundamentally different for super- and subsonic velocities. For supersonic velocity, the projectile leaves behind the effects (compression and dilation waves) generated at its previous positions, while for subsonic velocity it will be surrounded by them on all sides as by an atmosphere. Whoever has seen the beautiful photographs of these phenomena by E. and L. Mach will doubt the possibility of representing the multiplicity of the flow phenomena by a single law of air resistance, or even representing the different velocity regimes by a series of different laws that depend only on the instantaneous velocity of the center of gravity and the instantaneous position of the projectile about the center of gravity. The only adequate treatment of the problem, rather, consists in the simultaneous investigation of the motion of the fluid and the projectile by means of the hydrodynamic partial differential equations for the fluid and the dynamic total differential equations for the body, a treatment that would yield, in addition to the course of the motion, the exact law of air resistance.

In this generality, the problem is naturally completely unsolvable. One must subdivide it. A pure hydrodynamic problem is obtained if one imagines the motion of the projectile as given (for example, a uniform linear motion in the direction of its axis) and asks for the corresponding flow field. One can then find the energy dissipation of this flow field, or, somewhat more completely, the pressure that the flow field applies to the surface of the projectile. In the free motion of the projectile, this pressure, together with gravity, would successively change the assumed uniform and linear trajectory of the projectile. This too is an approximation, in so far as the changing velocity and position of the projectile would indeed influence, in reverse, the flow field and its pressure.

We continue, however, with the given simplest hydrodynamic problem. The energy dissipation in the entrained fluid may be divided into two parts, a first part in which the energy dissipation is due to waves, and a second part in which the energy transformation has its origin in friction or in eddies on the rear side of the projectile. To find the character of the first part, one begins from the differential equations of frictionless but compressible flow; to determine the second part, it suffices to adopt the differential equations of frictional or rotational but incompressible flow. The first part gives the "*wave resistance*" of the air; its presence is illustrated by the Mach photographs. The second part is concisely called the "*frictional resistance*" of the air. According to what was just said, it would be further distinguished into two components, the actual frictional resistance (surface resistance or skin friction) and the eddy resistance (form resistance, in the designation of L. Prandtl*)). These components may also be distinguished for the analogous problem of ship resistance.

With respect to the *frictional resistance*, it has been assumed since the time of Newton that this resistance is proportional to the square of the velocity, without, up to now, a successful proof of this on a hydrodynamic basis. The difficulty lies in the quadratic character of the hydrodynamic differential equations, the resulting instability of the simplest forms of motion, and their sudden change into complicated "turbulent" motions. This problem also appears rather hopeless in the case of projectile motion as long as it has not been possible to advance the hydrodynamic understanding of turbulence in the much simpler cases of flow in pipes, etc.

The problem of *wave resistance* may be simpler. In order to encourage the initiation of its investigation, we present some results from the analogous electrical case. If one moves an electric charge with a constant superluminal velocity, the resulting electric field may be determined (at least according to the old theory of the so-called stationary aether), and it shows the same features as the "Mach phenomenon" of hydrodynamics. The electrical charge (briefly speaking, the "electron")

*) Zeitschr. für Flugtechnik und Motorluftschiffahrt, Jahrg. 1, p. 63.

leaves behind its effects in a cone whose opening angle decreases with the velocity ratio v/c ($c =$ the velocity of light, $v > c$ is the velocity of the electron). In that effects of this kind are dissipated at infinity, a continuous energy loss by radiation occurs; a resistance opposite to the direction of its motion therefore acts on the electron. If one assumes that the electron is spherical, then the resistance becomes*)

$$A\left(1 - \frac{c^2}{v^2}\right);$$

the factor of proportionality A depends on the charge of the electron and the size of the sphere in a simple manner. The resistance is zero for $v = c$, and approaches a fixed limit for $v = \infty$. The field can also be determined for a subluminal velocity, and yields, in a known manner, zero resistance for arbitrary velocity $v < c$.

We have reason for the assumption that a similarly simple law for wave resistance also holds in the corresponding hydrodynamic problem of ballistics, in spite of conditions that are more complicated than those of electrodynamics. First, the complex aspherical shape of the projectile comes into consideration; second, the nonpenetration boundary condition on the surface of the projectile, which implies that the streamlines of the air must run parallel to the surface, is also much more difficult than the corresponding condition in electrodynamics, in which the divergence of the lines of electric force are determined directly by the magnitude of the charge density (the hydrodynamic problem is a "boundary value problem," while the electrodynamic problem is only a "summation problem"). The primary difference, however, lies in the more complicated nonlinear character of the hydrodynamic differential equations compared with the purely linear character of the electric field equations; this difference implies, for example, that electrical effects always and exactly propagate with the velocity of light c, while the speed of sound in hydrodynamics, which we may likewise denote by c, represents only a lower bound of propagation for infinitely small disturbances; the actual velocity of propagation increases with the amplitude of the disturbance. It thus follows that the region in front of the

*) A. Sommerfeld, Göttinger Nachtr. 1904, p. 401, and Amsterdamer Akademie, Proceedings, 1904, p. 366. Cf. also M. Abraham, Theorie der Elektrizität II, §27.[348]

electron is fully field-free in the electrodynamic case, while in the hydrodynamic case a compression wave lies in front of the projectile, and the surrounding surface of the Mach cone is formed from the compression wave toward the rear. This cushion of air will therefore be carried forward with a velocity greater than c; the speed of sound c is evidently no longer decisive for its propagation.

Our assumption that the electrodynamic law of wave resistance may nevertheless hold by and large for the hydrodynamic case is supported first by the reflection that the geometric character of the field is determined by the Mach cone in essentially the same manner for both cases, and second by the empirical results of firing tests. These results are represented by a curve that we take from the cited encyclopedia article by Cranz.

In Fig. 143, the velocity ratio $x = v/c$ is plotted as the abscissa, and the ordinate $y = (W + R)/R$ is the ratio of the total resistance, taken from ballistic tables, to the frictional resistance of the air, where we have decomposed the total resistance into the wave resistance W and the frictional resistance R; the latter is assumed to be proportional to v^2 according to the usual Newtonian law $R = av^2$. The solid curve represents an empirical interpolation formula of Siacci[349] that agrees very well with the extensive observational data. It is seen from the curve that y is constantly equal to 1 for small v; a rather sudden increase occurs in the vicinity of $v = c$, followed by a maximum and a further gradual decrease. We would explain this behavior theoretically in the following manner: for $v < c$, the wave resistance W is initially zero, and thus $y = 1$; in the vicinity of $v = c$ (and even somewhat before), the wave resistance begins to increase. It increases with v to a smaller degree, however, than the frictional resistance R; the ratio W/R thus decreases, as does y, with increasing v.

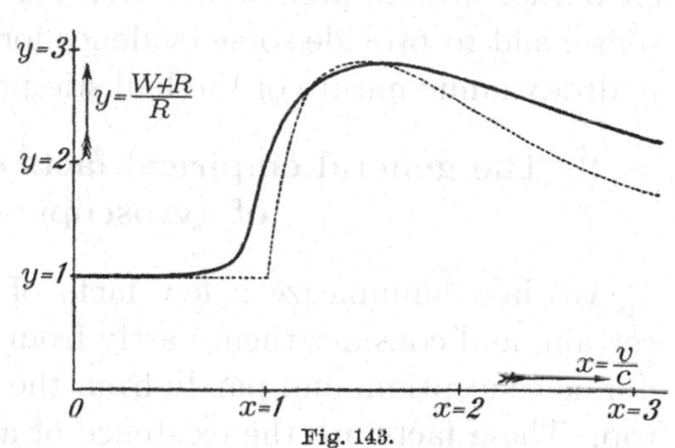

Fig. 143.

The dotted curve is the value of y that would result by direct transference of the above electrodynamic law for wave resistance. The naturally undetermined factor of proportionality A in the above formula is chosen so that the maximum of our electrodynamic law has the same value as the maximum of the empirical law. We therefore plot

$$W = 0, \; y = \frac{R}{R} = 1 \text{ for } v < c,$$

$$W = A\Big(1 - \frac{c^2}{v^2}\Big), \; R = av^2, \; y = 1 + C\frac{c^2}{v^2}\Big(1 - \frac{c^2}{v^2}\Big) \text{ for } v > c,$$

where $C = A/ac^2$ is obtained in the given manner from the observations. It can hardly be mistaken that we achieve, through this very simple point of view, a general agreement with the character of the empirical curve, which can scarcely be accidental.*)

For practical purposes, naturally, nothing is gained by our idealized form of the air resistance curve; its transmission is merely intended to encourage a more precise investigation of wave resistance in the above sense, and to provide some evidence for our hypothesis of the essential hydrodynamic nature of the ballistic problem.

9. The general empirical facts of ballistics and the role of gyroscopic effects.

We first summarize a few facts of ballistics that can be held as certain, and consider them partly from the standpoint of our hydrodynamic conception, and partly from the standpoint of the theory of the top. These facts are the existence of a moment due to air resistance, the sense of this moment, which corresponds to a point of application in front of the center of gravity of the projectile, the stability of a rotating long projectile, and the rightward deviation for right-handed rifling.

A i r r e s i s t a n c e a n d m o m e n t o f a i r r e s i s t a n c e. It is familiar to us that a moving body experiences a resistance from the surrounding medium. It is less well known, but clear from the preced-

*) According to a friendly communication from Mr. P r a n d t l, one can rigorously justify the Newtonian law of friction for very large velocities. The theoretical value of the factor of proportionality a is greater than the empirical value a_0 for small velocities. If one plots the ordinate of the figure as the total resistance divided by the frictional resistance for small velocities (that is, $y = (W + R)/a_0v^2$), with the retention of our electrodynamic value for W, then the right side of our dotted curve would be higher for large velocities by the amount $(a - a_0)/a_0$, because $R = av^2 > a_0v^2$. The agreement with the solid curve would thus be still better.

ing, that wave resistance makes up a predominant part of the total resistance under some conditions (in the vicinity of the speed of sound). If the body is formed symmetrically with respect to its direction of motion, the total resistance is naturally exactly opposite to the motion; the moment of the air resistance about the center of gravity vanishes. With the addition of gravity, however, this state of complete symmetry, which holds approximately at the moment of firing, is changed at each instant, since the tangent to the trajectory is turned downward. A moment due to air resistance would now be absent only if the air resistance were equivalent to a single force applied at the center of gravity. According to our hydrodynamic conception, this is naturally highly improbable; the complexity of the flow in the vicinity of the rapidly moving projectile produces a highly complicated distribution of pressure on its surface. Mach, for example, has measured a pressure decrease of more than one atmosphere over a distance of 13 mm near the tip of the projectile. The quantity that we wish to represent by the schematic concept of air resistance is determined physically, however, only by the totality of such pressure effects. We could not expect them to be equivalent to a single force at a special location (the center of gravity). Moments generally appear even in the much simpler case of a body in an incompressible and frictionless fluid, the only case that may be fully treated hydrodynamically. If we said above that the translational and rotational components of the motion are coupled in this case, this means nothing other than that the surrounding fluid set into motion influences the rotation of the body by a moment, and, conversely, that the pressure caused by the rotation influences the translational motion like a single-force. In the case of ballistics, the coupling of the two components of the motion is naturally not less intimate. The fiction of an empirically known air resistance should thus serve for the consideration of both components. We must, in any case, take account of their coupling, and give the air resistance a moment about the center of gravity.

The sense of the moment of air resistance. Application point of the resultant air resistance in front of the center of gravity. The sense of the moment of the air resistance is determined immediately by the known empirical fact that a projectile without artificial stabilization by rifling will tumble and become aligned perpendicular to its trajectory, in such a way that the tip of the projectile will point upward if it initially lay above the tangent to the trajectory, and downward if it lay below. The fictitious single-force that replaces the air resistance and is approximately opposite to the direction of motion thus has its point of application on the forward part of the projectile (as seen by the shooter behind the center of gravity).

This fact is well known from experience for slowly moving planar surfaces that are oriented obliquely with respect to their direction of motion. It forms, moreover, the basis for the theory of aviation. For slow motions (velocity small with respect to c), where it may be assumed that the air is incompressible, a theoretical justification can be given. The streamlines around an obliquely moving plate, calculated on the basis of ideal hydrodynamics, run in such a manner that they give a pressure distribution that is equivalent to a point force on the forward side of the plate; the eccentricity of the point of application increases as the angle between the direction of motion and the normal to the plate becomes larger. From the previously stated ballistic facts, we may assume that the same law remains valid for large velocities, in spite of the undoubtedly essential changes in the distribution of flow and pressure.

It has also been attempted to understand the facts in question from an entirely different standpoint. If wave resistance is completely disregarded and it is imagined that the frictional resistance is determined by Newton's formula av^2, where the normal component of the translational velocity for each surface element is used in the calculation as the effective velocity, one obtains a system of forces that is distributed over the surface. If these forces are composed according to the rules of statics, the point of application of the resultant lies in front of the center of gravity. In this calculation, however, only the pressure on the forward side of the projectile (the surface elements struck by the air flow) is considered; the simultaneously acting suction effects on the rear side, whose moments would act in the opposite sense, are arbitrarily neglected. We hardly need say, after the preceding, that we cannot regard either this consideration or that based on the assumption of incompressible flow as a fully valid proof of the relevant ballistic facts.

S t a b i l i z a t i o n o f t h e p r o j e c t i l e b y r o t a t i o n. One of the most well-known facts of ballistics is that a long projectile can be stabilized against tumbling by a rapid rotation about its figure axis. The rotational velocity produced by a rifled barrel is proportional to the outgoing velocity of the projectile; it amounts to more than 100 rotations per second for artillery weapons, and to more than 3000 rotations per second for infantry weapons. The long projectile (we think from now on of the artillery projectile) that is transformed in such a way into a *top* naturally no longer follows the moment of the air resistance, but rather deviates perpendicularly to it. While the external moment would generate a rotation about the transverse axis through the center of gravity for a nonrotating projectile, it merely causes a small change of position of the rotational impulse for a rapidly rotating projectile.

German guns have right-handed rifling; that is, each point moves in a right-handed screw. Seen from the front, therefore, the rotation occurs in the counterclockwise sense. If we wish to draw the rotational impulse on the figure axis to the front, which is recommended for clarity, then we must abandon our general rule for the positive sense of rotation, and regard a rotation in the counterclockwise sense as positive.

We first assume that the tangent to the flight path has momentarily descended with respect to the axis of the projectile. The moment of the air resistance then acts as an uprighting moment, and thus acts, seen from the right-hand side of the projectile, in the positive sense; its moment arrow (multiplied by the infinitely small time interval under consideration) is therefore applied to the right at the end of the eigenimpulse of the projectile (we naturally consider here only the rotational component of the impulse, the "impulse moment"), so that the eigenimpulse will be deflected to the right. If, in contrast, the axis of the projectile lies instantaneously below the tangent to the flight path, then the eigenimpulse will correspondingly be displaced to the left. In general, we can say that the endpoint of the impulse is displaced, because of the effect of the moment of the air resistance, about the tangent to the flight path in the same sense as the eigenrotation of the projectile.

To proceed further, we carry over the principal results of our theory of pseudoregular precession to the present case. We know that the figure axis always remains in the vicinity of the changing impulse axis for rapid rotation, and that the impulse axis, with neglect of the imperceptible nutations, moves, in the mean, on a cone about the vertical with constant mean precessional velocity P/N (cf. p. 303, equation (11)).

The cone always remains very narrow if the figure axis initially lay near the vertical. The vertical (that is, the direction of gravity) corresponds here to the direction of the fictitious resultant air resistance, to which the axis of the moment of the air resistance is perpendicular. The figure axis of the projectile will thus circulate, in the case of ballistics, about the resultant air resistance, which, in its turn, always falls approximately on the tangent to the flight path; the figure axis will indeed circulate in the direction in which the moment of the air resistance displaces the impulse (that is, in the direction of the rotation of the projectile), and the instantaneous motions of the impulse axis will be composed into a precession, or, as one says in ballistics, into a conical pendulum. The opening angle of the conical pendulum always remains very small, since the figure axis coincides with the tangent to the flight path (up to a small "outgoing error angle") at the moment of firing.

Our transference of the simple results for the top to the complicated circumstances of ballistics is naturally not without misgivings. The gravitational moment was $P \sin \vartheta$, where ϑ denoted the angle between the figure axis and the vertical. If we now denote by ϑ the corresponding angle between the axis of the projectile and the air resistance, we can write the assumption for the moment of the air resistance in the approximate form $P \sin \vartheta \cos \vartheta$. The latter vanishes not only for $\vartheta = 0$, on the basis of symmetry, but also for $\vartheta = \pi/2$, since the orientation of the projectile perpendicular to the flight path, which without rotation would be stable, gives a vanishing moment of air resistance. The factor P depends in an unknown manner on the velocity of the translation and that of the rotation. The analytic expression for the moment of air resistance is therefore different from that of the gravitational moment in the usual theory of the top, so that the dynamic effects in the two cases are also not exactly the same; one can remark that for small angles ϑ, the added factor $\cos \vartheta$ will be of small essential significance. More importantly, however, the tangent to the flight path, which indeed gives the approximate direction of the resultant air resistance, is not the spatially fixed vertical, but rather rotates downward because of gravity. For the preceding transference, however, we assumed that the figure axis of the projectile circulates about the changing direction of the air resistance with a small cone opening just as the figure axis of the top circulates about the fixed vertical for originally present approximate coincidence. The variability of the tangent to the flight path is by no means small; it can amount to 90° and more for a steep shot. The theoretical justification of our transference must show that the tangent to the flight path is to be regarded as slowly varying compared with the pendulum of the projectile, so that the figure axis has time for repeated circulations that adapt themselves to the changing position of the tangent to the trajectory. We wish, however, to refrain from any theoretical basis here, and simply call upon the undoubted empirical fact, confirmed by experts, that a projectile generally (that is, for not too steep an outgoing angle and for normal gun and projectile design) strikes with its tip (otherwise the explosion of the charge would indeed fail!), so that the axis of the projectile actually remains in the vicinity of the tangent to the flight path tangent or the resultant of the air resistance. This fact is obviously made possible only by the conical pendulum; that is, by the *occurrence of energetic gyroscopic effects.*

The rightward deviation of the projectile. The empirical facts are as follows: while the lateral deviation for infantry projectiles is small, there is a considerable rightward deviation for artillery projectiles with right-handed rifling; the deviation increases with increasing firing range as

firing range	500,	1000,	2000,	3000 m
rightward deviation	0,25,	1,1,	4,4,	11,5 m.

In modern weapons, moreover, the rightward deviation is automatically corrected by the placement of the gun sight.

It is very probable that we must view this phenomenon as a consequence of the conical pendulum, and thus an indirect consequence of the gyroscopic effect. In addition, it is known that the rightward deviation is transformed into a leftward deviation for left-handed rifling (Italian artillery has left-handed rifling). A precise quantitative theory of the rightward deviation fails on our ignorance of the law of air resistance, and could be derived only on a hydrodynamic basis. Even with the assumption of an empirical and certainly incomplete law of air resistance (a single force at an appropriately chosen point of application), the calculation is very detailed and possible only by successive approximations, because of the coupling between the translational and rotational components of the motion. We are thus satisfied with a qualitative estimation. We begin from the previously adopted representation that the axis of the projectile always circulates with constant mean angular velocity about the instantaneous position of the resultant air resistance in the sense of the eigenrotation. We assume that the tangent to the flight path and the resultant air resistance that approximately follows it descend with uniform velocity, which is permissible for not too long a time interval and under the preliminary neglect of the lateral deviation. If we now imagine a drawing plane placed perpendicular to the resultant air resistance at an arbitrary instant, then the intersection point O of the air resistance with this plane describes a downward directed line L with uniform velocity. Let the intersection point of the axis of the projectile with this same plane be Q. If we view the drawing plane from the projectile, then Q always moves perpendicularly to the line OQ with constant mean angular velocity in the clockwise sense. We thus have a most simple geometric image before us, from which the trajectory of Q follows immediately. The trajectory is evidently a *cycloid curve*, as is produced by the rolling of a wheel on a line. Point O corresponds to the instantaneous point of contact of the wheel, Q is a point on the wheel, and our straight line L is to be imagined as the rolling path. The rolling wheel lies, as seen from the projectile, *to the right* of L, since the wheel, just like the point Q fixed to it, rolls in the clockwise sense, and the contact point O should progress downward on L. According to whether Q lies on the circumference of the wheel, in its interior, or in its exterior, there results an ordinary cycloid with cusps, a oblate cycloid, or a prolate cycloid with loops. In fact, the above criterion that Q rotates with constant velocity about the instantaneous center of rotation of the wheel O is fulfilled for each cycloid curve. One may compare here, moreover, the entirely similar consideration of p. 295 for the treatment of pseudoregular precession in which cycloidal curves likewise appeared, and Fig. 48 there, which we must now imagine to be rotated in the clockwise sense by 90°, with the difference that the instantaneous center of rotation was given there by the instantaneous position of the impulse axis, while it is given here by the direction of the air resistance.

Whether an ordinary, oblate, or prolate cycloid is produced depends on the initial position of point Q with respect to point O, the precessional velocity of Q about O, and the downward velocity of point O, and remains undetermined for our consideration.*) Only one common

*) In the cited textbook of ballistics, Teil 1, Nr. 52 through 57, Mr. Cranz finds by an approximate solution that the trajectory is an ordinary cycloid whose arcs gradually broaden; this broadening, however, cannot occur in our representation, since we assume that the velocity with which tangent to the flight path descends (that is, the velocity of point O) is uniform. Other authors (cf. the literature references in Cranz, l. c.) are led by somewhat different calculations to prolate

feature of all three types is essential for us: *the trajectory of Q, as seen from the projectile, runs, in the mean, to the right of the line O*; the trajectories of the ordinary and oblate cycloids never cross to the left of O, and the loops of the prolate cycloid that cross to the left are shorter than the arcs on the right. Here, however, is an intuitive explanation of the rightward deviation: *since the apex of the projectile is found, in the mean, to the right of the vertical plane through the tangent to the flight path, the projectile will be pushed by the air resistance to the right.* The pressure on the left side of the projectile, namely, will predominate, so that a small component of the total air resistance acts directly to produce the lateral motion of the center of gravity, and the flight path will be deflected to the right.

This very superficial consideration can suffice as a qualitative explanation of the rightward deviation. That a quantitative treatment of the same fact is impossible, and that ballistic problems are still in a very unsatisfactory state, which is hardly to be denied,*) is due, as we wish to emphasize yet again, not to the theory of the top, which answers all questions posed to it in a specific and clear manner, but rather to hydrodynamics, with its much more complicated conditions and its infinitely increased possibilities of motion.

Until the middle of the 19th century, venerable mechanics held the undisputed leading place in both the theoretical conception of nature

or oblate cycloids. Cranz also points out (Nr. 57, l. c.) the uncertainty of the theoretical calculation and the necessity of systematic experiments.

*) Photographic images of projectile motion appear to be a promising means of assistance for experimental ballistics. F. Neesen,[350] on the one hand, has sought to record the conical pendulum on a photographic plate in the interior of the projectile, and the obtained curves are interpreted in the sense of the theory as cycloids of a very irregular character; on the other hand, a method has also been given to record the position of the projectile directly from the outside. Conclusive results have not yet been obtained for either technique because of unfavorable experimental conditions. Cf. Archiv f. Artillerie- und Ingenieuroffiziere, 53. Jg. 1889 and Kriegstechnische Zeitschrift 1903.

and the practical application of its power. In the last quarter of the 19th century, however, it must yield more and more to its younger rival, electricity. Theoretical electrodynamics leads more deeply into the understanding of nature than purely mechanical considerations, and electrical technology has opened more varied and more daring possibilities for energy utilization and transformation than the old mechanical constructions. It appears that mechanical technology in the 20th century will attack new problems with rejuvenated vigor by the application of the gyroscopic principle; it is not to be forgotten, however, that an important means of assistance, the actuation of the gyroscope, must be borrowed in most cases directly from the rival electrical technology. We evidently stand only at the beginning of this development; the possibilities offered here are still far from exhausted by the constructions presented in the preceding.

Translators' Notes.

275. (Advertisement) In September 1870, Maxwell gave an address in Liverpool to the Mathematical and Physical Sections of the British Association for the Advancement of Science [Maxwell 1870]. After some long and nebulous reflections on the nature of mathematics and physics in general, Maxwell turned to "the molecular theory of the constitution of bodies." In particular, he discussed Kelvin's molecular vortex theory:

> A theory, which Sir W. Thomson [later Baron Kelvin of Largs] has founded on Helmholtz's splendid hydrodynamical theorems, seeks for the properties of molecules in the ring-vortices of a uniform, frictionless, incompressible fluid. Such whirling rings may be seen when an experienced smoker sends out a dexterous puff of smoke into the still air, but a more evanescent phenomenon is difficult to conceive. This evanescence is owing to the viscosity of air; but Helmholtz has shewn that in a perfect fluid such a whirling ring, if once generated, would go on whirling for ever, would always consist of the very same portion of the fluid which was first set whirling, and could never be cut in two by any natural cause. The generation of a ring-vortex is of course equally beyond the power of natural causes, but once generated, it has the properties of individuality, permanence in quantity, and indestructibility. It is also the recipient of impulse and of energy, which is all we can affirm of matter; and these ring-vortices are capable of such varied connexions and knotted self-involutions, that the properties of differently knotted vortices must be as different as those of different kinds of molecules can be.
>
> If a theory of this kind should be found, after conquering the enormous mathematical difficulties of the subject, to represent in any degree the actual properties of molecules, it will stand in a very different scientific position from those theories of molecular action which are formed by investing the molecule with an arbitrary system of central forces invented expressly to account for the observed phenomena.

> In the vortex theory we have nothing arbitrary, no central forces or occult properties of any kind. We have nothing but matter and motion, and when the vortex is once started its properties are all determined from the original impetus, and no further assumptions are possible.

After this from Maxwell, physical theories based on the properties of rotating bodies must soon follow. In 1882, Joseph John Thomson (1856–1940) won the Adams Prize for a treatise in which he investigated the interaction of vortex rings, and "endeavored to apply some of the results to the vortex atom theory of matter" [Thomson 1883]. (Thomson later won the more memorable Nobel Prize for his discovery of the electron.) Kelvin constructed a gyroscopic model of the aether [Thomson 1890, pp. 466–472], and Sommerfeld published an early paper [Sommerfeld 1892] in which he used Kelvin's model to derive a system of equations for electromagnetic phenomena in nonconductors. A comprehensive review of vortex theories of the atom in nineteenth-century physics is given by Robert H. Silliman [Silliman 1963]. By the time that Vol. IV of the *Theorie des Kreisels* appeared in 1910, it was clear that these mechanical representations had no fundamental physical significance. But the persistent mental image of rotation reemerged when the designation "spin" was given to a property of subatomic particles.

276. (Advertisement) Fritz Alexander Ernst Noether (1884–1941) came from a family of mathematicians that included his sister Emmy Noether (1882–1935). The subject of his 1909 doctoral dissertation in Munich was the rolling of a sphere on a surface of revolution. His thesis advisor was Aurel Voss (1845–1931), but he also acknowledged the valuable inspiration of Sommerfeld [Noether 1985]. Noether taught applied mathematics and mechanics at Karlsruhe and Breslau until he was forced to retire in 1934. He then accepted an appointment at the University of Tomsk in the USSR. He was arrested by the People's Commissariat for Internal Affairs (NKVD) in 1937. No further information about his fate was available until 1989, when it was revealed that he had been executed in 1941 on a groundless accusation of anti-Soviet agitation [Parastaev 1989].

277. (Advertisement) Mr. cand. math. Behrens may be Wilhelm Behrens (1885–1917), who completed his doctoral degree under Felix Klein in 1911 with a dissertation (opus valde laudabile) entitled "*Ein der Theorie der Laval-Turbine entnommenes mechanisches Problem,*

behandelt mit Methoden der Himmelsmechanik" [Behrens 1911]. The problem treated in Behrens's dissertation is a nonlinear analysis of the motion of a rigid disk on a flexible shaft; a linearized analysis is given here in §8, no. 1. Behrens later became an assistant to David Hilbert (1862–1943). He died in 1917 while serving as a lieutenant in the German army.

278. (Advertisement) Hermann Franz Joseph Hubertus Maria Anschütz-Kämpfe (1872–1931) "belongs to the particularly mysterious and legendary class of inventors" [Broelmann 2002, p. 184]. The physicist Walther Gerlach (1889–1979) and Sommerfeld call him "the inventor of the gyroscopic compass," and write that "his life unfolds before us like a fairy tale, rich in trials and triumphs, begun under a favorable star, continued without decline in marital happiness, and ended with professional prominence" [Gerlach and Sommerfeld 1931]. Anschütz-Kämpfe spent his youth traveling, with occasional study of medicine and art history; his interest in the gyroscopic compass seems to have been motivated by his desire to reach the North Pole in a submarine. Anschütz-Kämpfe's long development of a practical gyroscopic instrument for navigation included a patent suit against the American Elmer Sperry (1860–1930); an expert opinion in the case was given by the former patent clerk Albert Einstein (see, for a start, [Broelmann 1998], [Lohmeier 2005], [Hughes 1971]). The company that Anschütz-Kämpfe founded in Kiel survives to this day as Raytheon Anschütz GmbH, a manufacturer of gyrocompasses, autopilots, and steering control systems.

Carl Julius Cranz (1858–1945) was professor of technical physics at the *Militärtechnische Akademie* and the *Technische Hochschule* in Berlin. From 1935 to 1937 he was scientific advisor to the Chinese government in Nanking. Cranz was an international authority on ballistics; the first two volumes of his *Lehrbuch der Ballistik* were translated into English by the United States Office of Scientific Research and Development [Cranz 1944].

Carl Diegel (1854–1931) was the chief torpedo engineer at the *Torpedowerkstatt* in Friedrichsort, now a suburb of Kiel. In 1903 he became director and technical advisor for buoys, torpedoes, and sea mines at Julius Pintsch AG, a firm originally founded for the manufacture of illuminating gas.

Ernst Otto Schlick (1840–1913) was an engineer and inventor who began his career in the shipbuilding industry by founding a shipyard for riverboats in Dresden. He eventually became director of Germanischer

Lloyd, a shipping classification society in Hamburg. In addition to his work on the gyroscopic ship stabilizer, Schlick developed a mass balancing system that allowed a significant increase in the size and speed of shipborne steam engines. A detailed obituary of Schlick is included in the *Jahrbuch der Schiffbautechnischen Gesellschaft* for 1914.

Maximilian Joseph Johannes Eduard Schuler (1882–1972) studied mechanics with August Föppl at the *Technische Hochschule* in Munich. From 1908 to 1922 he worked at the firm of his cousin Anschütz-Kämpfe in Kiel. Schuler was the director of the Institute for Applied Mechanics in Göttingen from 1934 to 1946.

Rudolf Skutsch (1870–1929) completed his habilitation in 1904 at the *Technische Hochschule* in Aachen, while Sommerfeld was a member of the faculty there. He was later professor of mechanics in Braunschweig and railroad construction inspector in Dortmund. In 1907, Skutsch received US patent 874,225 for a gyroscopic ship stabilizer that utilized two counterrotating flywheels (cf. p. 844).

Dieter Thoma (1881–1942) was an assistant to August Föppl from 1908 to 1910, and had previously worked for a manufacturer of turbines and regulators in Gotha [Broelmann 2002, p. 336]. He received the doctoral degree in 1910 with a thesis on the stability of water surge tanks in turbine plants, and later became director of the Hydraulic Institute in Munich.

279. (page 767) For the given sense of ϑ, the arrows that represent $N\,d\vartheta$ and K' in Fig. 116 should be reversed. This would explain the sign of K' in equation (IVe).

280. (page 771) Kelvin's gyrostat is shown in Fig. 204. It consists of a symmetric rotor that is concealed in a rigid casing. By mounting or suspending the casing in different ways, Kelvin constructs systems with two, three, or four degrees of freedom. Kelvin uses these different arrangements to illustrate the linear stability theory discussed here in §10, no. 3. The gyrostat "serves to illustrate the curious reversal of the ordinary laws of statical equilibrium due to the 'gyrostatic domination' of the interior invisible flywheel, when rotated rapidly" [Greenhill 1911].

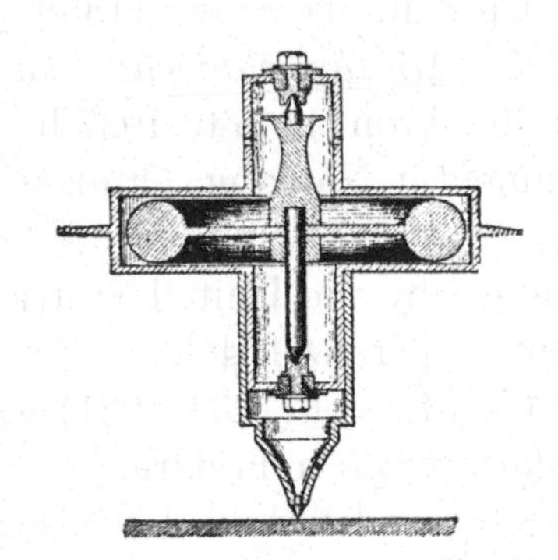
Fig. 204. Kelvin's gyrostat [Whitney 1889, p. 2699].

281. (page 771) We consider the problem posed by Sommerfeld in the following form. An axisymmetric rotor is mounted in a Cardanic suspension, as illustrated in Fig. 205. The outer ring of the suspension is subjected to an external moment $M(t)$ about the fixed vertical axis V. The orientation of the rotor is determined by the Euler angles φ, ψ, ϑ, defined with the left-handed rotation convention used by Klein and Sommerfeld. We neglect the masses of the inner and outer rings, air resistance, and bearing friction.

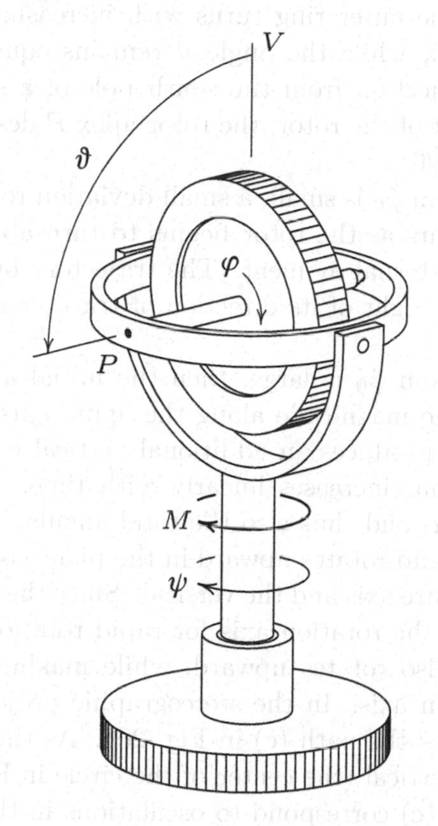

Fig. 205. Symmetric rotor with external moment $M(t)$.

We first consider the problem qualitatively, in the manner of Vol. II, Ch. IV, §§1 and 2. For this purpose, we assume that the angle ϑ has the initial value $\vartheta_0 = \pi/2$, the rotor has initial spin $\dot{\varphi}_0$, the inner and outer rings are initially at rest, and the moment M has a constant value M_0. We also assume a *mechanically spherical* rotor, for which the axial and equatorial moments of inertia are equal and the angular momentum vector is proportional to the angular velocity vector (cf. Vol. I, pp. 105–107). We will show below that the motion of the axisymmetric rotor can be obtained in a simple manner from that of the mechanically spherical rotor.

If the initial spin $\dot{\varphi}_0$ is zero, then the initial angular momentum of the rotor is also zero. The applied moment produces an angular momentum vector that is always directed along the vertical; the magnitude of this vector increases linearly with time. Since the angular momentum and the angular velocity are proportional for the mechanically spherical rotor, the outer ring turns with increasing angular velocity about the vertical, while the angle ϑ remains equal to $\pi/2$. In the stereographic projection from the south pole of a sphere centered at the fixed midpoint of the rotor, the rotor apex P describes the circular path (a) in Fig. 206.

If the initial spin $\dot{\varphi}_0$ is small, a small deviation resistance (cf. Vol. I, Ch. III, §5) appears as the rotor begins to turn about the vertical in response to the external moment. The trajectory of the rotor apex P is deflected to the right of its direction of travel, and follows the path (b) in Fig. 206.

If the initial spin $\dot{\varphi}_0$ is large, then the initial angular momentum is a vector of large magnitude along the figure axis of the rotor. The external moment produces an additional vertical component of angular momentum that increases linearly with time. The total angular momentum vector, and thus also the total angular velocity vector, increases in length and rotates upward in the plane containing the initial position of the figure axis and the vertical. Since the figure axis remains in the vicinity of the rotation axis for rapid rotation about the figure axis, the figure also rotates upward, while making small oscillations about the rotation axis. In the stereographic projection, the apex of the rotor describes the path (c) in Fig. 206. As the apex of the rotor approaches the vertical (the center of the circle in Fig. 206), the loops in the trajectory (c) correspond to oscillations in the precession angle ψ with increasingly large amplitude.

To investigate the motion of the rotor quantitatively, we begin with the Lagrange equations for the axisymmetric rotor, constructed according to the procedure of Vol. III, p. 558. The kinetic energy of the rotor is

$$T = \frac{A}{2}(\dot{\psi}^2 \sin^2 \vartheta + \dot{\vartheta}^2) + \frac{C}{2}(\dot{\varphi} + \dot{\psi} \cos \vartheta)^2,$$

where A and C are the equatorial and axial moments of inertia of the rotor, respectively. The work dW of the external moment for small

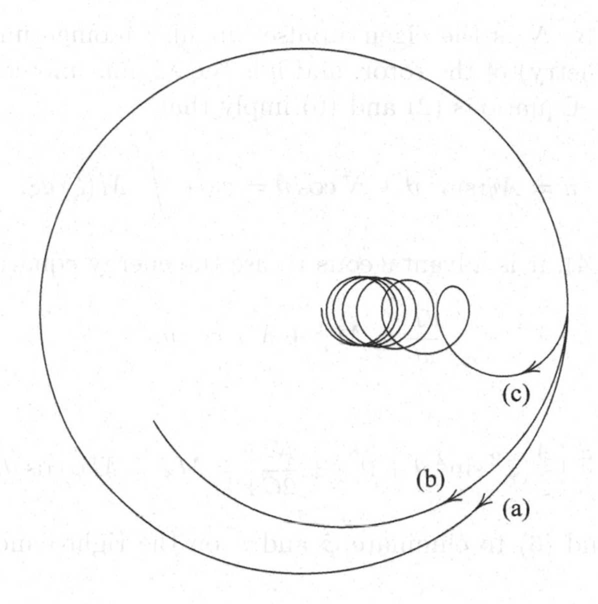

Fig. 206. Stereographic projections of the trajectories of the rotor apex P.

variations $d\varphi$, $d\psi$, $d\vartheta$ of the Euler angles is

$$dW = M\,d\psi + M\cos\vartheta\,d\varphi, \tag{1}$$

where we now assume that the moment M may depend on time. The Lagrange equations for the rotor are

$$\frac{dn}{dt} = M, \tag{2}$$

$$\frac{dN}{dt} = M\cos\vartheta, \tag{3}$$

$$A\frac{d^2\vartheta}{dt^2} + \frac{(N - n\cos\vartheta)(n - N\cos\vartheta)}{A\sin^3\vartheta} = 0, \tag{4}$$

where

$$N = \frac{\partial T}{\partial\dot{\varphi}} = C(\dot{\varphi} + \dot{\psi}\cos\vartheta), \tag{5}$$

$$n = \frac{\partial T}{\partial\dot{\psi}} = A\dot{\psi}\sin^2\vartheta + C\cos\vartheta(\dot{\varphi} + \dot{\psi}\cos\vartheta) = A\dot{\psi}\sin^2\vartheta + N\cos\vartheta. \tag{6}$$

As previously, N is the eigenimpulse (angular momentum about the axis of symmetry) of the rotor, and n is the angular momentum about the vertical. Equations (2) and (6) imply that

$$n = A\dot{\psi}\sin^2\vartheta + N\cos\vartheta = n_0 + \int_0^t M(\xi)\,d\xi. \tag{7}$$

In place of (4), it is advantageous to use the energy equation

$$\frac{dT}{dt} = M\dot{\varphi} + M\dot{\psi}\cos\vartheta,$$

or

$$\frac{d}{dt}\left[\frac{A}{2}(\dot{\varphi}^2\sin^2\vartheta + \dot{\vartheta}^2) + \frac{N^2}{2C}\right] = M\dot{\varphi} + M\dot{\psi}\cos\vartheta. \tag{8}$$

Using (5) and (6) to eliminate $\dot{\varphi}$ and $\dot{\psi}$ on the right-hand side of (8) gives

$$\frac{d}{dt}\left[\frac{A}{2}(\dot{\psi}^2\sin^2\vartheta + \dot{\vartheta}^2) + \frac{N^2}{2C}\right] = M\left[\frac{n}{A} + \left(\frac{1}{C} - \frac{1}{A}\right)N\cos\theta\right],$$

and use of equations (3) and (7) now gives

$$\frac{d}{dt}\left[\frac{A}{2}(\dot{\psi}^2\sin^2\vartheta + \dot{\vartheta}^2) + \frac{N^2}{2C}\right] = M\left(n_0 + \int_0^t M(\xi)\,d\xi\right) + \left(\frac{1}{C} - \frac{1}{A}\right)N\frac{dN}{dt},$$

which reduces to

$$\frac{d}{dt}\left[\frac{A}{2}(\dot{\psi}^2\sin^2\vartheta + \dot{\vartheta}^2) + \frac{N^2}{2A}\right] = M\left(n_0 + \int_0^t M(\xi)\,d\xi\right). \tag{9}$$

Equations (3), (7), and (9), grouped in the form

$$\begin{cases} \dfrac{dN}{dt} = M\cos\vartheta, \\ A\dot{\psi}\sin^2\vartheta + N\cos\vartheta = n_0 + \displaystyle\int_0^t M(\xi)\,d\xi, \\ \dfrac{d}{dt}\left[\dfrac{A}{2}(\dot{\psi}^2\sin^2\vartheta + \dot{\vartheta}^2) + \dfrac{N^2}{2A}\right] = M\left(n_0 + \displaystyle\int_0^t M(\xi)\,d\xi\right), \end{cases}$$

are three differential equations for the precession angle ψ, the inclination angle ϑ, and the eigenimpulse N. These three equations do not contain the axial moment of inertia C. The quantities ψ, ϑ, and N for the axisymmetric rotor can thus be found, for given initial values of these quantities, by assuming a mechanically spherical rotor with

$C = A$. In particular, the spatial trajectory of the axis of the rotor, which is determined by the angles ψ and ϑ, can be found by assuming a mechanically spherical rotor. When ψ, ϑ, and N have been found for the mechanically spherical rotor, the spin angle φ for the axisymmetric rotor can be determined from equation (5), which does involve the axial moment of inertia C. We thus have a result that is analogous to the Darboux theorem for the heavy top in Vol. II, Ch. IV, §5.

To complete the integration of the differential equations for the rotor, it is most convenient to assume a mechanically spherical rotor, and work in terms of the Cayley–Klein parameters α, β, γ, δ. The Cayley–Klein parameters are related to the Euler angles φ, ψ, ϑ by the equations

$$\begin{cases} \alpha = \cos\frac{\vartheta}{2}\cdot e^{\frac{i(\varphi+\psi)}{2}}, & \beta = i\sin\frac{\vartheta}{2}\cdot e^{\frac{i(-\varphi+\psi)}{2}}, \\ \gamma = i\sin\frac{\vartheta}{2}\cdot e^{\frac{i(\varphi-\psi)}{2}}, & \delta = \cos\frac{\vartheta}{2}\cdot e^{\frac{i(-\varphi-\psi)}{2}}, \end{cases}$$

$$\text{(10)} \qquad \begin{cases} \cos\vartheta = \alpha\delta + \beta\gamma, & \sin\vartheta = \sqrt{-4\alpha\beta\gamma\delta}, \\ e^{2i\psi} = \dfrac{\alpha\beta}{\gamma\delta}, & e^{2i\varphi} = \dfrac{\alpha\gamma}{\beta\delta}, \end{cases}$$

and satisfy the constraint equation

$$\text{(11)} \qquad \alpha\delta - \beta\gamma = 1.$$

The Lagrange equations for α, β, γ, δ are constructed according to the procedure in Vol. II, Ch. VI, §9. The kinetic energy is

$$T = 2A(\dot{\alpha}\dot{\delta} - \dot{\beta}\dot{\gamma}),$$

where A is the moment of inertia of the rotor, and the Lagrangian is

$$\mathcal{L} = T + \lambda(\alpha\delta - \beta\gamma),$$

where λ is a Lagrange multiplier. In terms of the Cayley–Klein parameters, the work dW given by (1) becomes

$$\begin{aligned} dW &= \frac{M}{2i}\left(\frac{d\alpha}{\alpha} + \frac{d\beta}{\beta} - \frac{d\delta}{\delta} - \frac{d\gamma}{\gamma}\right) + \\ &\quad + \frac{M}{2i}(\alpha\delta + \beta\gamma)\left(\frac{d\alpha}{\alpha} + \frac{d\gamma}{\gamma} - \frac{d\beta}{\beta} - \frac{d\delta}{\delta}\right) \\ \text{(12)} \qquad &= iM(\delta\, d\alpha - \gamma\, d\beta + \beta\, d\gamma - \alpha\, d\delta), \end{aligned}$$

where the constraint equation (11) has been used. The Lagrange equations for the Cayley–Klein parameters α, β, γ, δ become

$$2A\ddot{\alpha} - \lambda\alpha = iM\alpha, \tag{13}$$

$$2A\ddot{\beta} - \lambda\beta = iM\beta, \tag{14}$$

$$2A\ddot{\gamma} - \lambda\gamma = -iM\gamma, \tag{15}$$

$$2A\ddot{\delta} - \lambda\delta = -iM\delta. \tag{16}$$

What can be said of such beauty?

In order to solve equations (13)–(16), it is necessary to derive an expression for the Lagrange multiplier λ. This can be accomplished in three steps.

1. Multiplying (13) by δ, (14) by $-\gamma$, (15) by β, (16) by $-\alpha$, adding, and using (11) gives

$$2A(\ddot{\alpha}\delta - \ddot{\delta}\alpha + \ddot{\gamma}\beta - \ddot{\beta}\gamma) = 2iM,$$

$$\frac{d}{dt}\Big[A(\dot{\alpha}\delta - \dot{\delta}\alpha + \dot{\gamma}\beta - \dot{\beta}\gamma)\Big] = iM,$$

$$A(\dot{\alpha}\delta - \dot{\delta}\alpha + \dot{\gamma}\beta - \dot{\beta}\gamma) = \int_0^t iM(\xi)\,d\xi + in_0. \tag{17}$$

It can be verified that the left-hand side of (17) is i times the angular momentum n in equation (6) for our case $C = A$.

2. Multiplying (13) by $\dot{\delta}$, (14) by $-\dot{\gamma}$, (15) by $\dot{\beta}$, (16) by $-\dot{\alpha}$, adding, and using (17) and the time derivative of (11) gives

$$2A(\ddot{\alpha}\dot{\delta} + \ddot{\delta}\dot{\alpha} - \ddot{\gamma}\dot{\beta} - \ddot{\beta}\dot{\gamma}) = -iM(\dot{\alpha}\delta - \dot{\delta}\alpha + \dot{\gamma}\beta - \dot{\beta}\gamma),$$

$$\frac{d}{dt}\Big[2A(\dot{\alpha}\dot{\delta} - \dot{\beta}\dot{\gamma})\Big] = -\frac{iM}{A}\Big[\int_0^t M(\xi)\,d\xi + in_0\Big],$$

$$2A(\dot{\alpha}\dot{\delta} - \dot{\beta}\dot{\gamma}) = \frac{1}{A}\int_0^t M(\xi)\Big[\int_0^\xi M(\xi')\,d\xi' + n_0\Big]\,d\xi + E_0, \tag{18}$$

where E_0 is the initial kinetic energy of the rotor.

3. Multiplying (13) by $-\delta$, (14) by γ, (15) by β, (16) by $-\alpha$, adding, and using (11) gives

$$\lambda = A(\ddot{\alpha}\delta + \ddot{\delta}\alpha - \ddot{\beta}\beta - \ddot{\gamma}\gamma),$$

which after using the second time derivative of (11) becomes

$$\lambda = -2A(\dot{\alpha}\dot{\delta} - \dot{\beta}\dot{\gamma}).$$

Equation (18) now gives

$$\lambda = -\frac{1}{A}\int_0^t M(\xi)\Big[\int_0^\xi M(\xi')\,d\xi' + n_0\Big]\,d\xi + E_0, \tag{19}$$

which is an explicit expression for λ as a function of time.

With the Lagrange multiplier λ given by (19), equations (13)–(16) are second-order linear differential equations with known nonconstant coefficients. Various special cases for $M(t)$ can be considered. We pursue here only the case $M = M_0 =$ constant. In addition, we assume the initial conditions corresponding to the discussion of Fig. 205, so that $n_0 = 0$ and $E_0 = A\dot{\varphi}_0{}^2/2 = N_0^2/2A$. In this case, equation (13) for α becomes

$$\ddot{\alpha} + \Big(\frac{M_0^2}{4A^2}t^2 + \frac{N_0^2}{4A^2} - \frac{iM_0}{2A}\Big) = 0. \tag{20}$$

If we introduce the dimensionless time

$$\tau = \frac{M_0}{N_0}t,$$

then (20) becomes

$$\frac{d^2\alpha}{d\tau^2} + \Omega^2\Big(\tau^2 + 1 - \frac{i}{\Omega}\Big)\alpha = 0, \tag{21}$$

where the dimensionless parameter Ω is defined by

$$\Omega = \frac{N_0^2}{2M_0A}.$$

The equation for β is identical to that for α, and the equations for γ and δ are obtained from those for α and β merely by substituting $+i$ for $-i$. It is easy to solve these equations numerically, and thus obtain a numerical solution for the Euler angles ψ and ϑ by means of equations (10). Fig. 206 was produced from such numerical solutions; the trajectory (b) corresponds to $\Omega = 0.1$, and the trajectory (c) corresponds to $\Omega = 3$. Equation (21) and the analogous equations for β, γ, and δ can also be solved analytically in terms of parabolic cylinder functions of complex order and argument [Magnus 1954, pp. 91–94]. Asymptotic properties of the parabolic cylinder functions can then be used to derive the (interesting but not practically important) large-time behavior of the rotor.

When Ω is large (the case of rapid initial spin), it is possible to derive simple approximations for the motion of the rotor. For large Ω, a first-order WKB solution of equation (21) is

$$\alpha = \frac{1}{r^{1/4}}\Big[C_1 \exp\Big(i\Omega \int r^{1/2}\, d\tau\Big) + C_2 \exp\Big(-i\Omega \int r^{1/2}\, d\tau\Big)\Big],$$

where

$$r = \tau^2 + 1 - \frac{i}{\Omega}$$

and C_1 and C_2 depend on initial conditions. Analogous approximations can be made for β, γ, and δ. Expanding these approximations to first order in $1/\Omega$, we find, for the initial conditions assumed here,

$$\begin{aligned}\cos\vartheta &= \alpha\delta + \beta\gamma \\ &= \frac{\tau}{\sqrt{\tau^2+1}} - \frac{1}{2\Omega}\sin\Omega\phi(\tau), \end{aligned} \tag{22}$$

where

$$\phi(\tau) = \tau\sqrt{\tau^2+1} + \log(\tau + \sqrt{\tau^2+1}). \tag{23}$$

The form of (22) has been encountered many times before. When the initial spin is large, the apex of the top rises slowly toward the vertical; the slow elevation is accompanied by small and rapid oscillations. For small τ, (22) and (23) give

$$\begin{aligned}\cos\vartheta &= \tau - \frac{1}{2\Omega}\sin 2\Omega\tau, \\ \vartheta &= \frac{\pi}{2} - \tau + \frac{1}{2\Omega}\sin\Omega\tau \\ &= \frac{\pi}{2} - \frac{M_0}{N_0}t + \frac{AM_0}{N_0^2}\sin\frac{N_0}{A}t,\end{aligned}$$

which is Sommerfeld's approximation (IVg) on p. 769.

282. (page 773) Arthur Mason Worthington (1852–1916) was headmaster and professor of physics at the Royal Naval Engineering College in Devonport, England. His primary field was fluid mechanics, especially the study of splashes. In the introduction to the sixth edition of his book on rigid body rotation, Worthington admits to having introduced the slug as "the British Engineer's Unit of Mass" [Worthington 1920, p. viii].

283. (page 774) Kötter's paper in the *Sitzungsberichte der Berliner Mathematischen Gesellschaft* is included in an appendix to Vol. 7 of the *Archiv der Mathematik und Physik* [Kötter 1904]. The sittings for 1903–1904 include an unpleasant exchange between Kötter and Carl Cranz over an article by Kötter on projectile motion.

284. (page 778) An English translation of the report of the *Studiengesellschaft für elektrische Schnellbahnen* (Society for the Study of High-Speed Electric Trains) was published in 1905 [Welz 1905]. An English summary of plans for the trial runs is given in *Engineering* by Oskar Lasche (1868–1923), who directed the construction of the locomotive that attained a speed of 210 km/hr [Lasche 1901].

285. (page 781) The mechanical engineer Gustav Wittfeld (1855–1923) worked for the Prussian state railway in Bromberg, Cassel, Frankfurt, Cologne, and Berlin. He received an honorary doctorate in 1917 from the *Technische Hochschule* in Charlottenburg for his contributions to the electrification of the German railway system. In 1923, stricken by poor health, he committed suicide, "a truly courageous act after the example of the heroes of classical Greece" [Wechmann 1923].

286. (page 781) Gustav Brecht (1880–1965) was an engineer who directed the electric railway division of the *Allgemeine Elektricitäts-Gesellschaft* from 1907–1911. He became chairman of a coal mining company in Cologne, and was a member of the board of directors of the *Reichsverband der deutschen Industrie.* In the cited paper, Brecht refers to a "much discussed" derailment of an electrically powered train on the New York Central Railway in February 1907. Twenty people were killed when a passenger train derailed in the Bronx, two days after the introduction of electrified rail service. Brecht considers two properties of electric locomotives that may have contributed to the crash: the low position of the center of gravity, and the gyroscopic effects of the rotating masses. After a long and thorough comparison, Brecht concludes that "a correctly constructed electric locomotive is undoubtedly superior to a steam locomotive with respect to capability for smooth operation and high velocity of travel" [Brecht 1910].

287. (page 782) Robert Whitehead (1823–1905) studied engineering in Manchester, England, but spent his professional career in France, Italy, and Austria. He was the manager of a foundry in Fiume when he met the Austrian naval officer Giovanni Biagio Luppis von Rammer (1813–1875), who had conceived a self-propelled floating device, laden with explosives and controlled from the shore by ropes, that could be directed toward ships near the coastline. Whitehead developed the idea into a practical weapon, formed a company for its manufacture, made sales and license agreements around the world, and became a wealthy man. Whitehead's torpedo works at Fiume is described and illustrated in *Engineering*, Vol. 72, 1901, pp. 398–401.

288. (page 784) The depth-control apparatus of the Whitehead torpedo is shown in Fig. 207. The compartment **C** of the engine room is flooded, so that hydrostatic pressure is applied to the left on the rubber diaphragm **b** of the hydrostatic piston. The pendulum bob **v**, supported by the two pendulum rods **v'**, swings in the plane of the lower drawing. The motion of the hydrostatic piston is transmitted to the control valve **f** by the hydrostatic piston lever **g**, the upper connecting rod and sleeve **h**, the cross lever **i**, and the lower connecting rod and sleeve **l**. The rock-shaft **m** transmits the motion of the system of levers through the bulkhead by means of the air- and watertight stuffing box **n**, which is bolted to the forward side of the bulkhead against a washer of thick paper. A plug in the top of the stuffing box screws in against a porpoise-hide washer. The desired running depth of the torpedo is set by turning the adjusting screw **p**, which changes the compression in the hydrostatic piston spring **c** by raising or lowering the upper arm of the bell crank **e**.

289. (page 784) Lieutenant Commander Walter J. Sears (1858–1913) was an inspector at the Bliss Torpedo Works in Brooklyn, New York, the company that manufactured the Whitehead torpedo for the United States Navy. His terse paper in *Engineering* [Sears 1898], which includes a two-page photographic plate, is taken from the *Proceedings of the United States Naval Institute*, Vol. XXIV, No. 1, 1898, pp. 89–110. Sears describes the Whitehead torpedo as "nearly in the shape of a porpoise." He also gives a detailed description of the many adjustments that must be made before the torpedo can operate correctly.

290. (page 784) Ludwig Obry (1852–1942) graduated in 1870 from the nine-year *Oberrealschule* in Görz (Gorizia, Italy) and learned shipbuilding in Trieste, which was then an important port of the Austrian empire. In 1885 he became a first-class design draftsman (a salaried employee (*Gastig*) without rank) in the Austrian navy [k. u. k. Kriegsmarine 1893, p. 57]. He was granted US patent 562,235 (1896) for his torpedo guidance apparatus. Subsequent US patents 621,364 (1899) and 648,878 (1900) were granted for improvements in the starting device and the rotor bearings.

Obry also invented a gyroscopic device for firing ships' guns at the instant when one or both of the main axes of the ship are horizontal, in order "to render the flight of projectiles discharged from ships' guns independent of the rolling or lurching motion of the vessel" (US patents 595,820 (1897) and 680,066 (1901)).

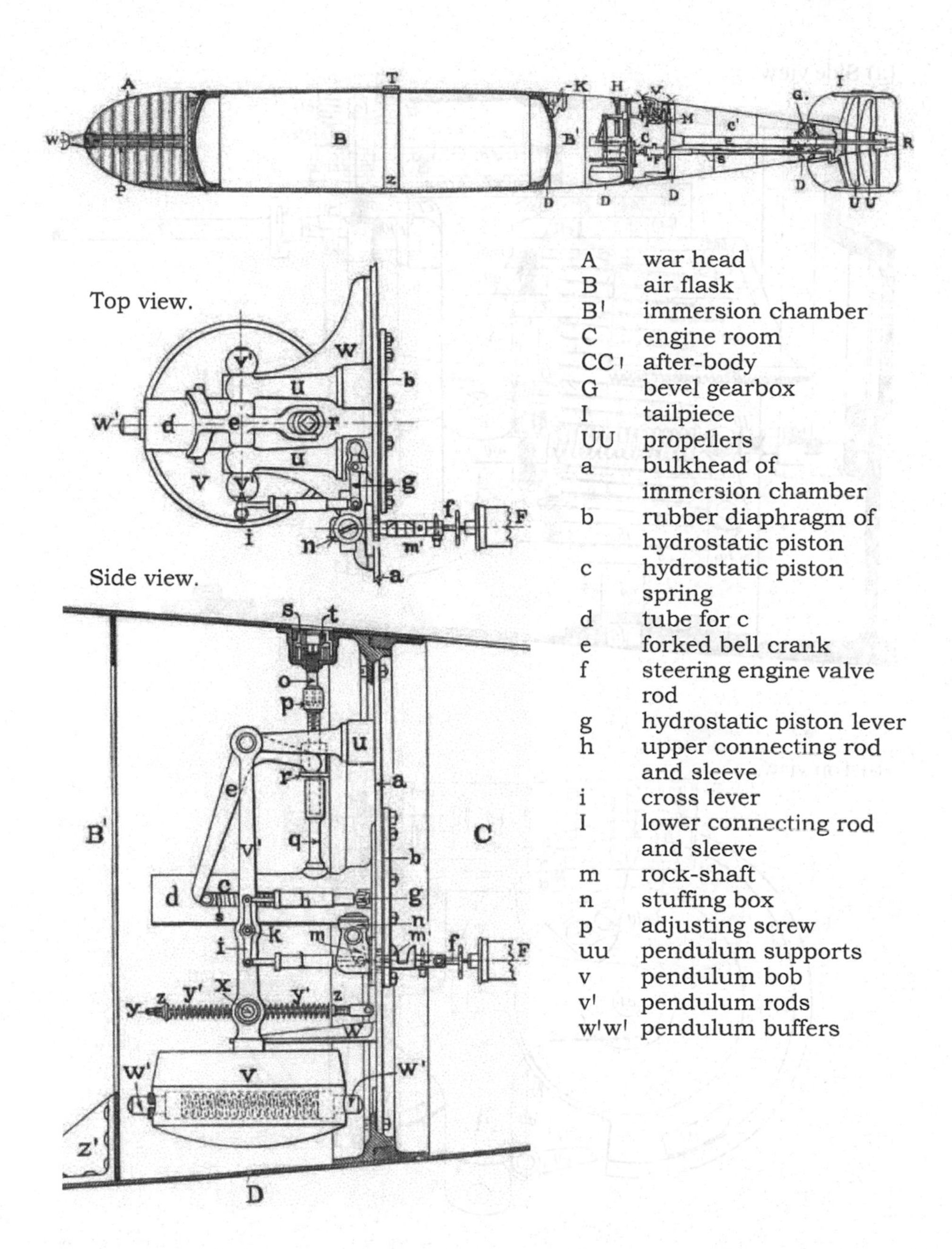

Fig. 207. Depth-control apparatus of the Whitehead torpedo [Bureau of Ordnance 1898].

(a) Side view.

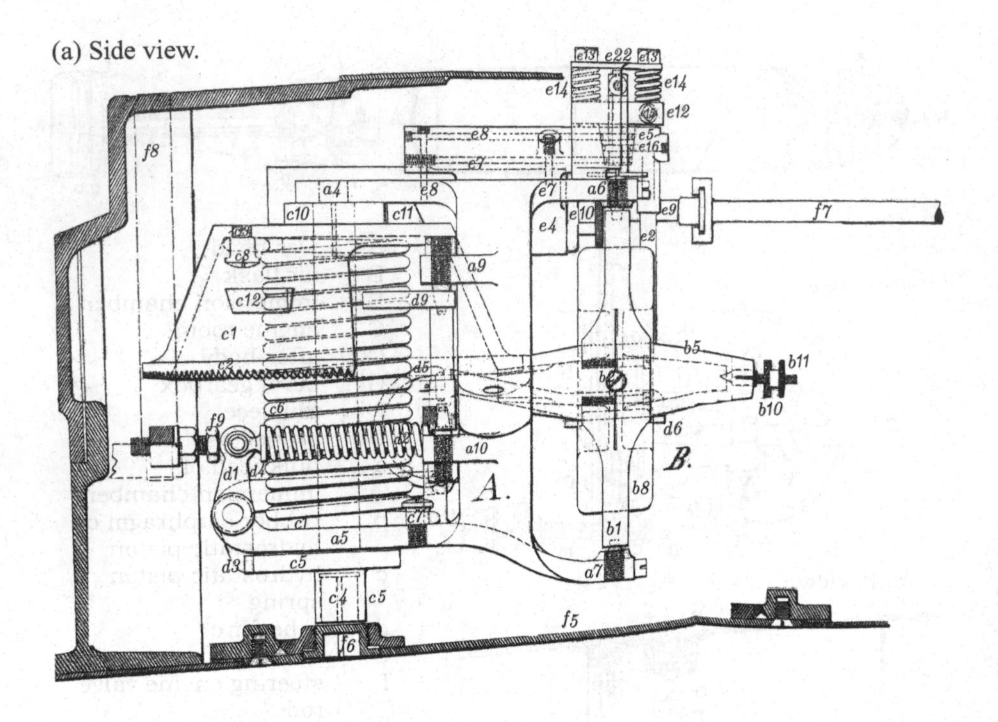

(b) Top view.

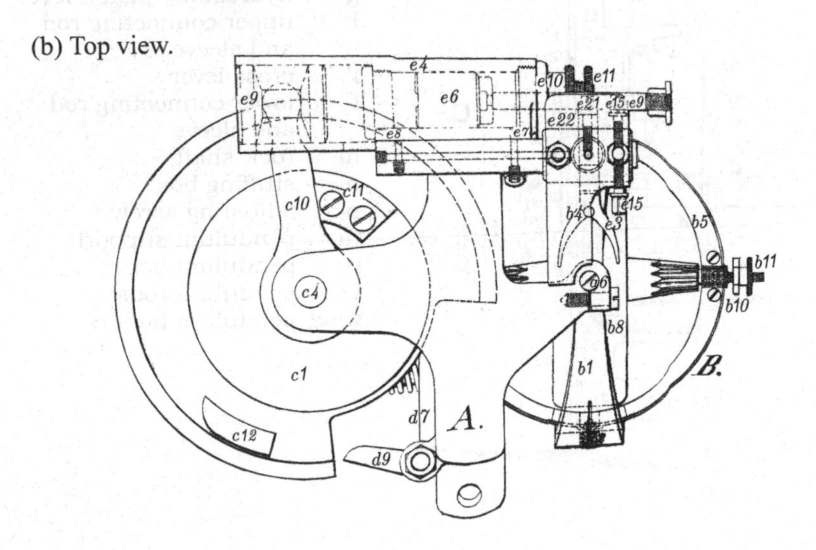

Fig. 208. Guidance apparatus of the Whitehead torpedo.

(c) Rear view.

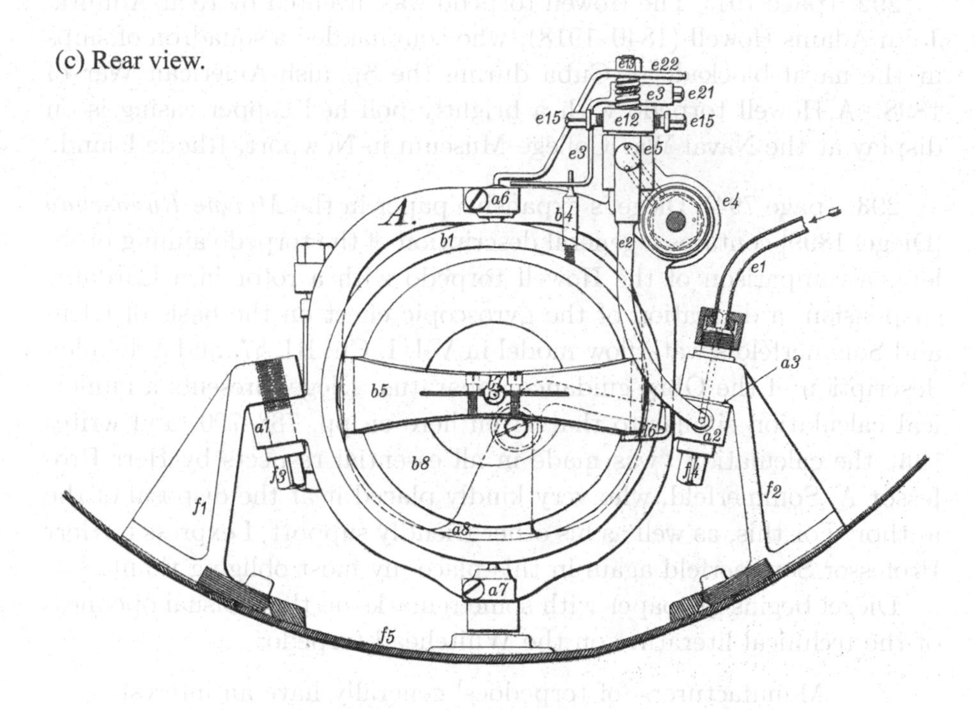

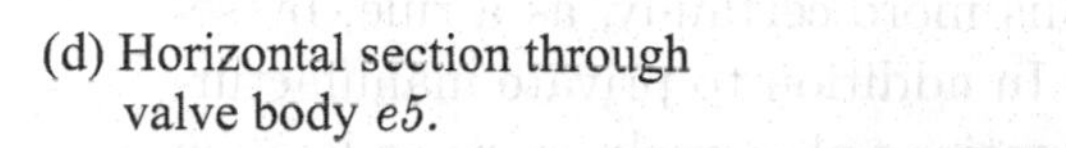
(d) Horizontal section through valve body *e5*.

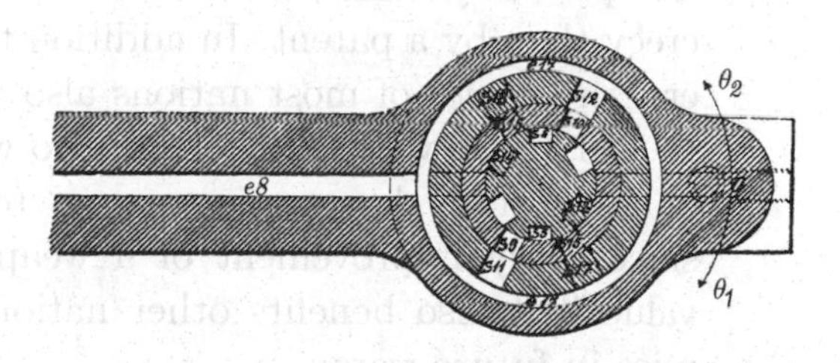

291. (page 786) Some details of the guidance apparatus for the Whitehead torpedo are shown in Fig. 208 [Diegel 1899]. The dowel on the inner ring in Fig. 121 of p. 785 is labeled *b4* in Fig. 208. When the body of the torpedo turns, this dowel causes a rotation of the fork *e3*, which aligns the pressurized air channels in the valve body *e5*. The rotor *b8* is spun by the toothed drum *c1*, which is tightened before firing against the stiff coil spring *c6*.

292. (page 791) The Howell torpedo was invented by Rear Admiral John Adams Howell (1840–1918), who commanded a squadron of ships in the naval blockade of Cuba during the Spanish-American War of 1898. A Howell torpedo with a brightly polished copper casing is on display at the Naval War College Museum in Newport, Rhode Island.

293. (page 794) Diegel's expansive paper in the *Marine-Rundschau* [Diegel 1899] contains a general description of the torpedo aiming problem, a comparison of the Howell torpedo with a rotor in a Cardanic suspension, a derivation of the gyroscopic effect on the basis of Klein and Sommerfeld's waterflow model in Vol. I, Ch. III, §7, and a detailed description of the Obry guidance apparatus. Diegel presents a numerical calculation similar to that given here on pp. 788–790, and writes that the calculation "was made in all essential respects by Herr Professor A. Sommerfeld, who very kindly placed it at the disposal of the author. For this, as well as his other friendly support, I express to Herr Professor Sommerfeld again in this place my most obliging thanks."

Diegel begins his paper with some remarks on the unusual openness of the technical literature on the Whitehead torpedo:

> Manufacturers [of torpedoes] generally have an interest to preserve new designs and improvements as their own, and probably achieve this more certainly, as a rule, by secrecy than by a patent. In addition to private manufacturers, the navies of most nations also work, more or less, on the further development of torpedo weapons. These navies only naturally have a reason not to publish innovations, since every improvement of a weapon loses considerable value if it also benefits other nations who might be enemies in future wars.
>
> The recently much discussed O b r y apparatus for the directional guidance of torpedoes appears to be an exception in this regard. It is not only patented in almost all cultured nations, but the drawings and description of the apparatus have already been published by Mr. W. J. S e a r s, lieutenant in the United States Navy, in the Proceedings of the United States Naval Institute, Vol. XXIV, No. 1. "Engineering" then reprinted this article on page 89 of Vol. LXVI, No. 1698, July 15, 1898. The apparatus has thus become publicly well known.

294. (page 794) An article in the Sunday magazine of the New York Times for Sept. 24, 1897, gives a description of public trials of the Whitehead torpedo in Sag Harbor, New York:

> The torpedoes were tested most thoroughly. When the torpedo boat Porter was at the proving grounds, three buoys, thirty yards apart, were anchored in the bay. The first range was 1,000 yards. The vessel was started at full speed, and when it reached the firing points discharged the torpedo at the center of the buoy. All of the shots were effective. It was so with the 1,200, 1,500, and 2,400 yard ranges. Logs and other obstructions were put in the path of the torpedo. When it stuck one of them it would deflect, but the gyroscopic attachment brought the head of the torpedo back to a course parallel with the original and within a few feet of the path it was fired into. The tests from the floats were severe, but of the most satisfactory nature.

295. (page 794) Sir Henry Bessemer (1813–1898) devoted a chapter of his interesting autobiography [Bessemer 1905] to his unsuccessful attempt to build a steamship with a stabilized saloon cabin. The chapter opens with Bessemer's motive for this effort:

> Few persons have suffered more severely than I have from seasickness, and on a return voyage from Calais to Dover in the year 1868, the illness commencing at sea continued with great severity during my journey by rail to London, and for twelve hours after my arrival there. My doctor saw with apprehension the state I was in. He remained with me throughout the whole night, and eventually found it necessary to administer small doses of prussic acid, which gradually produced the desired effect, and I slowly recovered from this severe attack. My attention thus became forcibly directed to the causes of this painful malady, which I, in common with most other persons, attributed to the diaphragm being subjected to the sudden motions of the ship. Hence, as a natural sequence, its cure appeared only to require that some mechanical means should be devised whereby that part of the ship occupied by passengers should be so far isolated as to prevent it from partaking of the general rolling and pitching motions. In this way I

> entered, almost without knowing it, into an investigation of the subject; and gradually, as my ideas were developed, I determined to make a model vessel, small enough to be placed on a table, and to which the usual pitching motion of a ship was imparted by clockwork.

A fanciful early conception of Bessemer's stabilized saloon cabin is shown in Fig. 209. Although his stabilization system was a failure, Bessemer considered the saloon cabin itself to be a great success:

> It formed a room 70 ft. long by 30 ft. wide, with a ceiling 20 ft. from the floor; its beautiful morocco-covered seats; its fine carved-oak divisions and spiral columns; its gilt, moulded panels, with hand-painted cartoons; its groined ceiling, tastefully decorated, gave an idea of luxury to the future Channel passage which all seemed to appreciate.

296. (page 794) The cited article by the English engineer and inventor Beauchamp Tower (1845–1904) describes a gyroscopically stabilized platform that Tower constructed as a mount for shipborne guns [Tower 1889]. An obituary notice for Tower in the *Minutes of Proceedings of the Institution of Civil Engineers* for 1905 states that the English Admiralty "rejected the apparatus on account of the extra weight involved, which it was thought might be more profitably utilized. Much discouraged, Mr. Tower nevertheless proceeded to adapt the platform for passenger-seats on cross-channel steamers, and he was engaged on this work at the time of his death, which occurred suddenly from cerebral hæmorrhage, at his residence, Hillstead, Brentwood, Essex, on the 31st December, 1904." During the discussion of Tower's paper, John Macfarlane Gray (1832–1908) took the opportunity to point out that he had recognized Bessemer's error in 1874, designed and patented a gyroscopic stabilizer with the necessary three degrees of freedom, and demonstrated a working model of his device, "which gave so much satisfaction that the original designer made, with me, a preliminary arrangement for applying my method to the already constructed swinging saloon. This intention was not carried out, but that was not due to any question about the soundness of the plan I proposed."

297. (page 795) In 1904, Schlick presented his design for a gyroscopic ship stabilizer, with working models, to the British Royal Institution of Naval Architects, of which Schlick was a member [Schlick 1904].

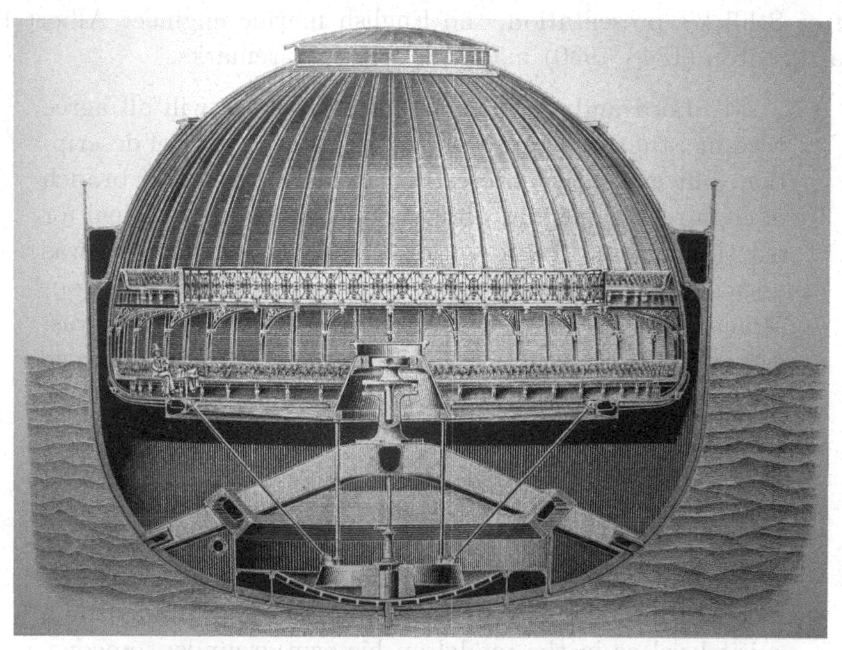

SECTION THROUGH EARLY FORM OF BESSEMER SALOON, IN STILL WATER

SECTION THROUGH EARLY FORM OF BESSEMER SALOON, WITH VESSEL ROLLING

Fig. 209. Early conception of Bessemer's stabilized saloon cabin [Bessemer 1905, plate XXXVII].

After Schlick's presentation, the English marine engineer Albert Edward Seaton (1848–1930) made the following remarks:

> My Lord and Gentlemen, I am sure you will all agree with me, that this paper is not only of a very novel description, but extremely interesting; in fact, it is a new branch of engineering entirely, and has opened up a great deal for our thought and consideration. Thirty-one years ago I was associated with Sir Edward Reed in the design and construction of the then celebrated S.S. *Bessemer.* She was, as you know, specially designed and built for the passage between Dover and Calais. She had a saloon amidships, hung on bearings, that was supposed to keep quite steady in any weather; the reason of my mentioning that unfortunate ship to-night is that, strange to say, the saloon was to be kept in the horizontal position by means of machinery controlled by a gyroscope. At the outset, Sir Henry Bessemer had intended the control to be by hand, guided by a spirit level as in the model on his own grounds; someone, however, whispered "gyroscope" to him, and talked a great deal about it. Eventually a scheme was devised whereby the gyroscope was to be used for keeping the saloon in a level position. Unfortunately, Sir Henry's adviser—I have forgotten now who he was—was not quite so well up in the knowledge of the gyroscope as he might have been, and he assumed that it had more energy than it really possesses. It came to the ears of my old friend Mr. Macfarlane Gray, who wrote to the papers and exploded the whole thing, much to the disgust and disappointment of Sir Henry ... Mr. Gray will speak for himself, and show how the fallacy of the gyroscope was exposed on that occasion. The application of it to the purposes of keeping ships steady in the way Herr Schlick suggests is, however, work of a different nature.

Macfarlane Gray then gave an account of his part in Bessemer's project and made some comments on Schlick's design:

> My Lord and Gentlemen ... In 1874, Sir Henry Bessemer's gyroscopic controller for the swinging saloon of the *Bessemer* steamship was described in *Engineering.* An adverse criticism by me of this design appeared in that journal for October 16, 1874, on page 307, and, in consequence, it

was abandoned. The axis of the trunnions was in Bessemer's design fixed fore and aft, through a misunderstanding of the principle of the gyroscope. At the same time I patented the arrangement of axes which is exhibited in Herr Schlick's design. ... Sir Henry Bessemer would have fitted his gyroscope with the axes as in Herr Schlick's arrangement, and according to my patent, but his co-directors and he differed about the amount he should contribute to the expense; and so the plan I had amended was likewise dropped. The invention before us differs from Bessemer's also in another way. Herr Schlick proposes to directly restrain the rolling of the hull by the gyroscope. In Bessemer's design the gyroscope only controlled a hydrostatic valve, and thereby the swinging was to be automatically controlled by hydraulic rams. A comparatively small gyroscope would suffice to move a valve, while a very large one might be insufficient to restrain the rolling of a steamship directly. There is no doubt, however, about its being possible to create a steadying force by Herr Schlick's method, with a proportionally massive gyroscope. It would be very instructive to see this invention carried out practically. The proposal is to mount in a passenger steamer a 10-ton flywheel, the rim to have at its circumference a velocity of 447 miles an hour, or just one-fourth of the muzzle velocity of the 12-inch projectiles in the *Duncan*, given in Sir E. J. Reed's paper yesterday. At this velocity a hoop of steel of any diameter and one square inch in cross section has a bursting tension of 20 tons on the square inch. In Germany, this installation might perhaps be allowed in passenger steamers, but I am afraid our Board of Trade surveyors would not pass it.

The seemingly lively discussion of Schlick's paper also included polite disagreements about gyroscopic effects for paddle steamers and bicycles. Schlick concluded the discussion as follows:

> My Lord and Gentlemen, I very much regret that I have not sufficient command of your language to be able to reply to the various points which have been raised in the discussion. I shall be much pleased, however, to send full replies in writing so that these may be added to the Transactions.

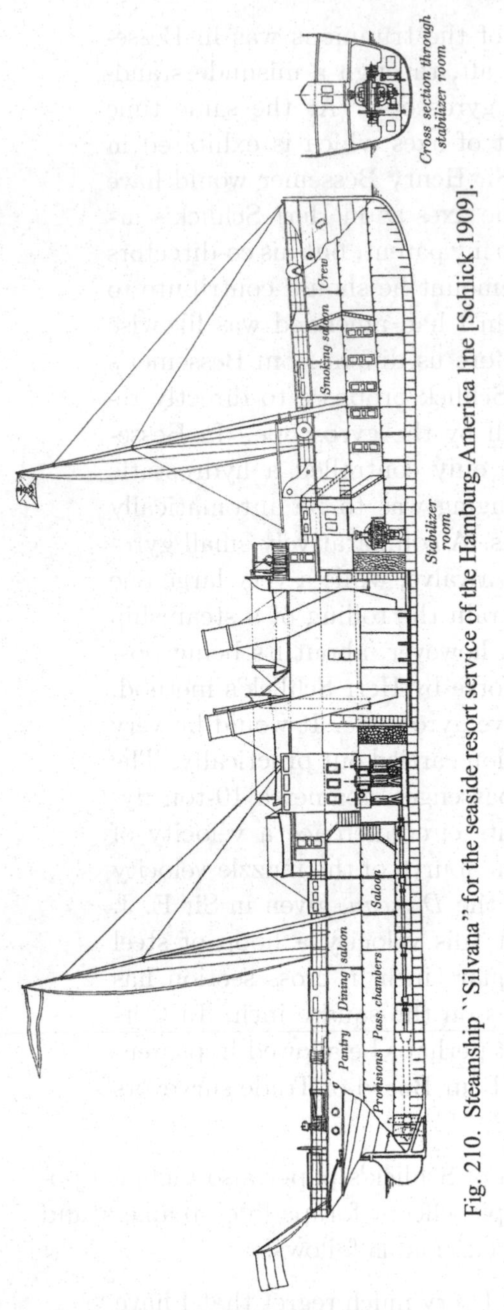

Fig. 210. Steamship "Silvana" for the seaside resort service of the Hamburg-America line [Schlick 1909].

I should like to emphasize once more, that my principle object was to raise interest in the subject, from the purely scientific point of view, leaving the question of its practicability to a more convenient period. I am afraid many of you, gentlemen, did not see very much of the experiments I made this evening; but if any of you are interested in this matter, I will repeat the experiments after the close of this meeting. In conclusion, allow me to express my sincere thanks for the kind manner in which you have received my paper.

298. (page 795) A longitudinal section of the steamship "Silvana" is shown in Fig. 210. The length between perpendiculars is 62.58 m, and the width of the frames is 8.99 m. The stabilizer is remarkably small in proportion to the ship.

Hermann Föttinger was the chief designer in the shipyard AG Vulcan (or Vulkan) in Stettin. In 1909 he became professor and director of the Institute for Fluid Dynamics at the *Technische Hochschule* in Danzig, and in 1924 he became head of the department of fluid physics and turbomachinery at the *Technische Hochschule* in Berlin-Charlottenburg. Föttinger's gyroscopic stabilizer for the "Silvana" is shown in Fig. 211. The large piston and cylinder near the front are part of the hydraulic system for damping the oscillations of the stabilizer frame.

Fig. 211. Gyroscopic ship stabilizer on the steamship "Silvana" [Schlick 1909].

299. (page 798) The thirty-ninth section of Stäckel's encyclopedia article "The elementary [!] dynamics of systems of particles and rigid bodies" is a historical review of the dynamic theory for rigid bodies that float in a fluid [Stäckel 1898]. Stäckel attributes the concept of the metacenter and the corresponding stability theory to Pierre Bouguer (1698–1758), Leonhard Euler (1707–1783), and Charles Dupin (1784–1873).

300. (page 798) The resistance of a ship to rolling motion is considered in section 4 of the encyclopedia article "The theory of the ship," by Alexei Nikolaevich Krylov (1863–1945) and Conrad Heinrich Müller (1878–1953) [Krylov 1898].

301. (page 805) Krylov and Müller attribute the given theory of forced rolling motion to the English engineer and naval architect William Froude (1810–1879) [Krylov 1898].

302. (page 808) Lorenz writes that "Herr Ingenieur O. Schlick of Hamburg called the attention of the author, some time ago, to the peculiar effects, analogous to strong damping, of a top on the rolling motion of a model ship. Since this phenomenon, to my knowledge, has not yet been treated in the literature, the present attempt at a theory of the same may perhaps claim some interest" [Lorenz 1904]. Schlick's demonstration model of the gyroscopic ship stabilizer (perhaps the same seen by Lorenz) is shown in Fig. 212. The underside of the model is formed by circular arcs whose centers lie at the metacentric height of the model cross section. Schlick does not give dimensions of the model, but states that it rocks "like a cradle" on a tabletop.

Fig. 212. Demonstration model of the gyroscopic ship stabilizer [Schlick 1909].

303. (page 808) After obtaining favorable experimental results with his model, Schlick asked Föppl to make calculations to determine whether similarly favorable results could be expected for a full-size implementation on a ship. Föppl writes that his calculations "confirmed

in all essential points the conclusions that the inventor had reached through experiments with his model, and supplemented them to the extent that one can now proceed with all confidence to the realization of plans on a larger scale. This work is now in process" [Föppl 1904a].

Föppl also comments on the previous work of Lorenz:

> As I had just finished the editing of this article for publication, I received from Prof. Lorenz in Göttingen an offprint of his paper "*Die Wirkung eines Kreisels auf die Rollbewegung von Schiffen*" in the *Physikalische Zeitschrift*, 1904, p. 27. The calculations of Mr. Lorenz agree in many points with the theory presented here, as is not to be otherwise expected. The main difference is that the Lorenz formulas are developed only for the case of no damping, and therefore $k = 0$. It has been proved here, however, how important damping is for the practical utility of the stabilizer. Also, Lorenz's work does not contain completely executed numerical calculations. Thus my work is not made superfluous by the earlier work of Lorenz, and I see no reason, with consideration of the calculations of Mr. Lorenz, which I accept as completely correct within their domain of validity, to make any change to my article.

304. (page 808) Paul Rurik Bruno Malmström (1872–1919) received his doctoral degree in physics in Leipzig; he studied from 1896–1897 in Göttingen, and lists Klein and Sommerfeld among his teachers there [Malmström 1905]. Malmström later became professor of physics in the University of Helsinki. In his seventeen-page pamphlet *Die Theorie des Schlick'schen Schiffskreisels* (which is subtitled "I.", but does not appear to have been continued), Malmström cites the earlier work of Lorenz and Föppl, and writes that "since these works, however, consider only the free (damped or undamped) oscillations in still water, I have also developed, at the friendly suggestion of Prof. A. Sommerfeld, the theory for the case in which the motion of the ship is determined by an external periodic force" [Malmström 1909].

Föppl was not impressed by Malmström's work. In the sixth volume of Föppl's *Vorlesungen über technische Mechanik*, Malmström receives the following footnote [Föppl 1910, p. 272]:

> I mention here that Mr. R. Malmström has published a treatise on the theory of the gyroscopic ship stabilizer in the Acta Societatis scientiarum Fennicae, Vol. 35, Hel-

singfors, 1907, in which he considers the case that the ship is subjected to regularly recurring lateral waves. It is assumed in the major part of this work that the frame of the rotor is not braked. At the conclusion, however, the braked frame is also treated briefly. The free oscillations are not introduced, but only the forced oscillations that are added to them. I have not been able to convince myself that there is to be seen in this work a significant advance for practice compared with my previously given theory, but rather believe that my treatment in its presently given scope should be retained, as long as a change is not suggested by experiments.

305. (page 841) The Austrian marine engineer and reservist Franz Berger (b. 1873) makes two rather strong criticisms of the gyroscopic ship stabilizer: (1) it is superfluous, because rolling motion of a ship can be reduced much more simply by a bilge keel, and vertical rather than horizontal motion is the primary cause of seasickness; (2) it is dangerous, because it converts rolling motion to pitching motion, and its large internal energy cannot be quickly dissipated in case of a disaster at sea [Berger 1904]. In a reply to Berger, Föppl addresses the coupling between rolling and pitching motions with the arguments given here on pp. 841–844. Föppl does not attempt to address the first criticism, saying that he "is not able to answer the given objections to the gyroscopic stabilizer from a practical standpoint, since I cannot consider myself competent to judge whether it is generally worthwhile to make such expenditures to suppress rolling motion as the application of this means requires" [Föppl 1904b].

306. (page 845) The passive gyroscopic stabilizer developed by Schlick fell out of use after the death of its inventor. Elmer Sperry, who met met Schlick in Germany in 1909 and conducted experiments on the "Silvana," invented an active gyroscopic stabilizer that was installed on the US naval destroyer *Worden* and the Italian ocean liner *Conte di Savoia* [Hughes 1971, pp. 103–128; de Santis 1936]. The gyroscopic stabilizer has been superseded for large ships by active fin stabilizers, but gyroscopic stabilizers are still manufactured by the Mitsubishi Corp. for use in luxury yachts.

307. (page 846) Edmond Paulin Dubois (1822–1891) was professor of hydrography in the *École Navale* in Brest. The cited paper in

the *Comptes Rendus* [Dubois 1884] is a report on sea trials with the *gyroscope marin*, one of several navigational instruments invented by Dubois. The *gyroscope marin* is not a compass, but rather a gyroscopic device intended to indicate a fixed spatial direction for a few minutes' time, thus allowing an accurate measurement of the turning angle of a ship during a change of course [Lucas 1918, pp. 41–42]. In addition to his maritime research, Dubois published an annotated French translation of Gauss's *Theoria motus corporum coelestium* [Dubois 1864].

308. (page 846) Thomson's paper on the gyrostatic model of the magnetic compass is reprinted in Vol. IV of his *Mathematical and Physical Papers* [Thomson 1910]. Thomson mentions the idea of floating the rotor of a gyroscopic compass, but describes in detail an impractical arrangement in which the rotor is suspended from "a very fine steel bearing wire, not less than 5 or 10 metres long (the longer the better; the loftiest sufficiently sheltered enclosure conveniently available should be chosen for the experiment)." He also suggests that "the locality be anywhere except at the North or South pole."

309. (page 846) The development of the gyroscopic compass is described and analyzed in depth by Jobst Broelmann in his *Intuition und Wissenschaft in der Kreiseltechnik*, the best reference for the historical context of all the devices discussed in Vol. IV of the *Theorie des Kreisels.* The gyroscopic compass patented by Marinus Gerardus van den Bos and Barend Janse was developed primarily by Leendert Janse (1818–1898), a professor in the maritime college in Amsterdam. Barend Janse, a cousin of Leendert and an administrative officer in the Dutch Navy, participated as a financial sponsor. Van den Bos was a clergyman who practiced instrument making as a hobby [Broelmann 2002, p. 83].

310. (page 846) Heinrich Meldau (1866–1937) studied mathematics in Göttingen and became professor in the maritime college in Bremen. Meldau devotes forty-eight pages of his article in the *Encyklopädie der mathematischen Wissenschaften* to the problem of using a magnetic compass on board an iron ship [Meldau 1906].

311. (page 846) Narziß Ach (1871–1946) was a physician, inventor, and experimental psychologist. His interest in gyroscopic devices stemmed partly from their possible relevance to investigations of human perception and cognition, and partly from his personal character as a "man of many natures" [Broelmann 2002, pp. 170–183]. Ach was present when Anschütz-Kaempfe spoke before the *Schiffbautechnische*

Gesellschaft, and made the following remarks during the discussion period [Anschütz-Kaempfe 1909]:

> Your Royal Highness! Gentlemen! The positioning of the meridian-gyrocompass in the north–south direction follows, as we have just heard, from the rotation of the Earth. The first to point this out was Foucault, more than fifty years ago. The gyroscopic compass obviously reacts, however, not only to the motion of the Earth, but also to the motion of the ship; its starts, stops, changes of course, etc., all influence the placement of the rotation axis. The credit for first indicating this belongs to Herr Dr. Martienssen of the firm of Siemens & Halske.
>
> The essential feature of the apparatus of Herr Dr. Anschütz-Kaempfe appears to me to be his damping mechanism for the reduction of these deviations from the north–south direction. At the same time, it appears to me that with the introduction of this damping mechanism, a certain defect of the apparatus is also present. As you recall, this damping mechanism includes a pendulum whose relative displacement with respect to the rotation axis initiates a turning-moment that seeks to return the rotation axis to the north–south direction. I have rejected such a construction on various grounds.
>
> You will not be surprised if I inform you that the solution to the problem of the gyroscopic compass has been attempted energetically from other sides for a long time. I myself have been occupied with the solution of this problem for several years. The firm of Hartmann & Braun in Frankfurt a. M. has undertaken the implementation and the practical design of this apparatus, and the nautical department of the Imperial Admiralty has supported us in the most commendable manner.
>
> Our most recent constructions of the meridian-gyrocompass have the advantage that they almost completely avoid the introduction of a damping mechanism; further, the rotational moment of the rotor can be much smaller, so that we can therefore use a reduced rotation number and a smaller weight of the rotating mass.
>
> Unfortunately, I am not able to enter into the construction of this apparatus in more detail today. I intend, how-

> ever, to make a thorough report to you on the subject at the next general meeting.
>
> I only remark further that the solution of the problem of the gyroscopic compass also permits us to address other problems that are of significance for shipping practice, and that the gyroscopic compass will then find lasting application in the merchant marine as well. Of this too I beg leave to report to you at the next general meeting. (Applause.)

Ach never presented the design of his gyroscopic compass to the *Schiffbautechnische Gesellschaft*, and his association with Hartmann & Braun ended in 1911. A single gyroscopic azimuth keeper by Hartmann & Braun survives at the Deutsches Museum in Munich [Broelmann 1998].

312. (page 848) The physicist Oscar Martienssen (1874–1957) "thoroughly examined the problem of constructing a gyroscopic compass on behalf of the firm of Siemens & Halske" [Martienssen 1906]. Martienssen concluded his paper with the following summary:

> It follows from this theoretical-experimental investigation that for installation on fixed ground, the directional force that a rotor experiences due to the effect of the Earth's rotation can very well be used for the construction of a rotational compass. The construction can be chosen without essential practical difficulty so that the directional force of such a compass, and therefore also the certainty of the positioning, will be essentially stronger than that of a magnetic compass, and the oscillation period will be not essentially greater than that of the magnetic compass.
>
> Such a compass, however, is unusable on a vehicle, since it will be disturbed by its motions. In order that this disturbance remain sufficiently small, the oscillation period must be greatly enlarged by an appropriate construction, and indeed an oscillation period of at least 30 minutes will be necessary for a large vehicle (warship, etc.); for a small vehicle (torpedo boat, submarine), a still essentially larger oscillation period is required. Then, however, the general application of such a slowly swinging compass may not be possible, at least as a steering compass, since incidental oscillations of the compass needle cannot be distinguished from motions of the ship.

Anschütz-Kaempfe was well aware of Martienssen's work, and proceeded to make his compass usable on ships by inventing a damping mechanism that depended on the airflow generated by the spinning rotor. The German Imperial Navy adopted Anschütz-Kaempfe's compass in 1908, but Martienssen, who also attended Anschütz-Kaempfe's presentation to the *Schiffbautechnische Gesellschaft*, was not convinced:

> Royal Highness! Gentlemen! I am happy to learn from the communication of Herr Anschütz that he has succeeded in constructing a steering compass according to the gyroscopic principle, which, due to a very long oscillation period, is not disturbed, within practical bounds, by changes in the velocity of the ship.
>
> In this respect, the apparatus corresponds precisely to the theory that I published two years ago for the gyroscopic compass, which I could then study and test only with an experimental model.
>
> I believe, however, that there still now remain the same doubts I expressed with respect to the practical applicability of such a compass, and that it is indeed not possible, for such long oscillations of the compass rose, to distinguish incidental oscillations from motions of the ship.
>
> If we assume that the compass rose is set by some unlucky chance into oscillation, these oscillations will persist for one or two hours, and the helmsman will set an entirely false course, since he has not the possibility to distinguish, for a long time, whether or not such an oscillation has occurred.
>
> The occurrence of large disturbances has indeed become less probable by means of the ingenious method of damping implemented by Herr Anschütz, since incidental rhythmic repetitions of disturbing causes do not cause a strengthening of the deviation from the north–south direction. I believe, however, that sufficient possibilities of disturbance remain for such a sensitive apparatus, so I may place before Herr Anschütz the same question that was asked of me two years ago by the Imperial Navy:
>
> Can you guarantee for this instrument with complete certainty that for a specific time, say only 24 hours, a deviation greater than 3° from the north–south direction will not occur?

I believe that the well-known experiment presented here proves only what was also claimed by me, that such a disturbance will not occur for a long time under favorable circumstances, and it appears to me now, as it did then, that the introduction of the gyroscopic compass into the navy is possible only under such a restriction of the requirements.

The presently introduced form of the compass is applicable to the merchant marine in still other ways. As Herr Prof. Schilling stated, it may be of use not as a steering compass, but rather as a control compass for the magnetic compass. The long oscillation period, which is entailed only by the motion of the ship, is then unnecessary, since such control measurements can be made when the ship is nearly stationary. A gyroscopic compass with an oscillation period of 1/2 to 1 1/2 minutes is now feasible without essential difficulties; such a gyroscopic compass would be a very convenient and reliable control device. I thus believe, in contrast to Herr Prof. Schilling, that the gyroscopic compass is also of significance for the merchant marine in this implementation.

I may take this occasion to remark that the English, French, and Dutch Navies conducted extensive experiments with the gyroscopic compass in the 1880s. The compass itself did its duty for long voyages in these experiments, especially in the French Navy, but did not attain adoption. An apparatus that Herr van den Bos constructed for the Dutch Navy in 1887 is available for inspection in the presentation rooms of the firm Siemens & Halske by those gentlemen who take part in tomorrow's tour to the *Wernerwerk*.*

Martienssen left Siemens & Halske in 1910, was employed briefly by Anschütz-Kaempfe, started his own company that produced gyroscopic navigation instruments for nautical and mining applications, and eventually become professor of technical physics in the University of Kiel. In 1923 (just before Max Schuler's important paper discussed in note 316), Martienssen published a nicely illustrated review of the history and then current state of the gyroscopic compass [Martienssen 1923].

*A factory of the firm founded by Ernst Werner Siemens (1816–1892; after 1888, von Siemens) and Johann Georg Halske (1814–1890) in Siemensstadt, a locality in the Spandau district of Berlin.

313. (page 848) A schematic representation of the gyroscopic compass described on p. 848 is shown in Fig. 213. The line SN is the horizontal south–north direction, and V is the vertical. The angles ϑ and ψ are defined as on p. 853: ϑ is the elevation angle of the figure axis F with respect to the horizontal and thus also the angle between the plane of the rotor and the vertical, and ψ is the rotation angle of the outer ring, measured to the west of the position in which the figure axis is in the meridional plane.

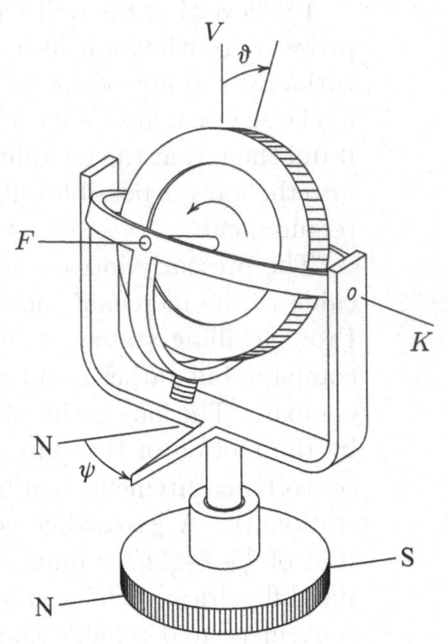

Fig. 213. Schematic representation of the gyroscopic compass.

314. (page 854) The angular velocity components and the gyroscopic effects for the rotor of the gyroscopic compass are shown in Fig. 214. The component $\omega \cos \varphi$ of the Earth's rotation about the horizontal south–north direction has components $\omega \cos \varphi \sin \psi$ about the line of nodes OK and $\omega \cos \varphi \cos \psi$ about the projection of the figure axis F into the horizontal plane. The gyroscopic effects $N\, d\vartheta/dt$ and $N\omega \cos \varphi \sin \psi$ act about the line perpendicular to OF and OK; their components about the vertical are obtained by multiplication by $\cos \vartheta$.

315. (page 862) An English-language description of the Anschütz gyroscopic compass as it existed in 1910 is available in a small book published by Sir George Keith Buller-Fullerton-Elphinstone (1865–1941), chairman of Elliott Brothers Ltd., the British licensing partner of Anschütz & Co. in Kiel [Elphinstone 1910]. The book includes a theoretical chapter with calculations by Schuler, rearranged by Harold Crabtree and Alfred Lodge (1854–1937) "so as to accord with English mathematical symbols and practice."

316. (page 863) The gyroscopic compass error caused by the rolling motion of a ship remained a problem for many years. Attempts to solve

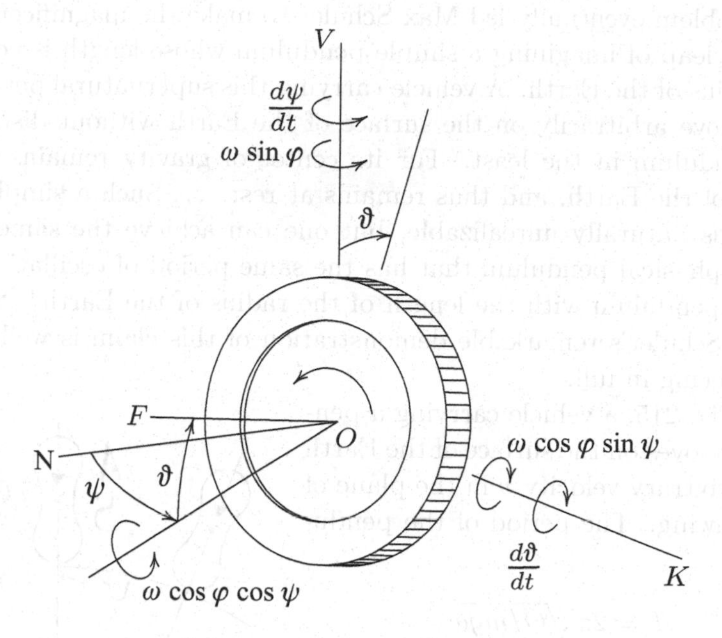

(a) Angular velocity components.

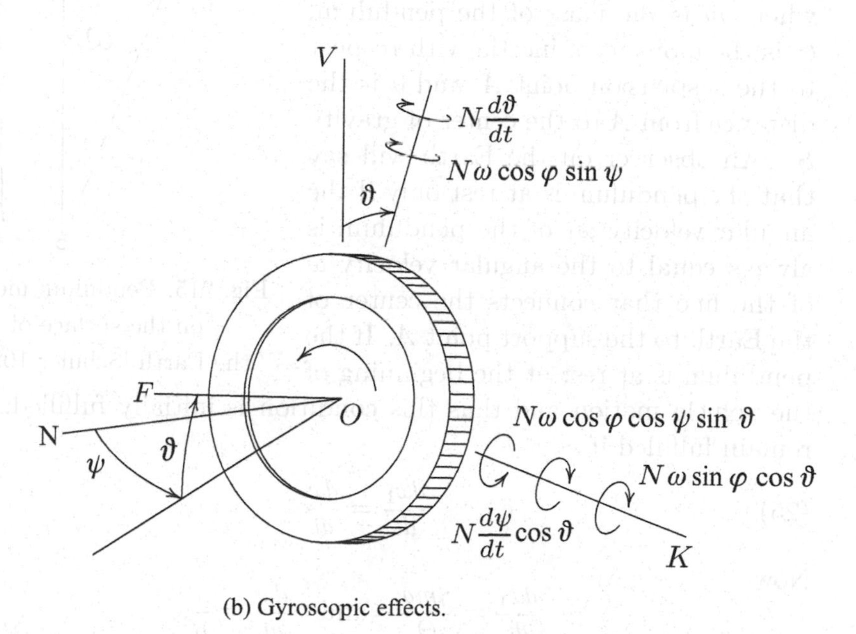

(b) Gyroscopic effects.

Fig. 214. Angular velocity components and gyroscopic effects for the rotor of the gyroscopic compass.

this problem eventually led Max Schuler to make the magnificent conceptual leap of imagining a simple pendulum whose length is equal to the radius of the Earth. A vehicle carrying this supernatural pendulum "can move arbitrarily on the surface of the Earth without disturbing the pendulum in the least. For its center of gravity remains at the center of the Earth, and thus remains at rest. ... Such a simple pendulum is naturally unrealizable, but one can achieve the same effect with a physical pendulum that has the same period of oscillation as a simple pendulum with the length of the radius of the Earth" [Schuler 1923]. Schuler's remarkable demonstration of this claim is well worth reproducing in full.

In Fig. 215, a vehicle carrying a pendulum moves on the surface of the Earth with arbitrary velocity v in the plane of the drawing. The period of the pendulum is

$$T = 2\pi\sqrt{\Theta/mga}, \tag{24}$$

where m is the mass of the pendulum, Θ is the moment of inertia with respect to the suspension point A, and a is the distance from A to the center of gravity S. An observer on the Earth will say that the pendulum is at rest only if the angular velocity ω_1 of the pendulum is always equal to the angular velocity ω of the line that connects the center of the Earth to the support point A. If the pendulum is at rest at the beginning of the vehicle motion and thus this condition is initially fulfilled, it will remain fulfilled if

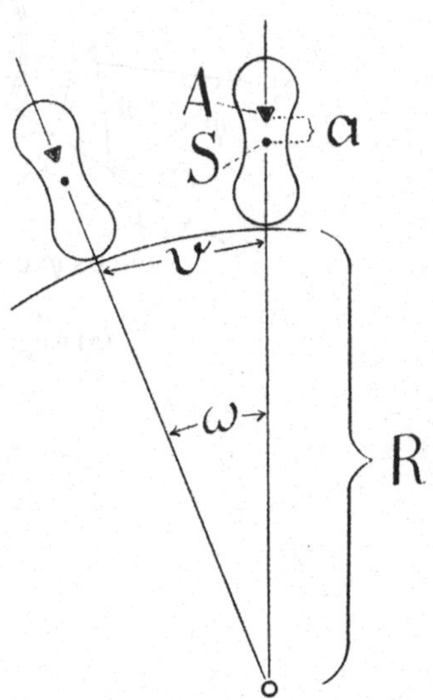

Fig. 215. Pendulum moving on the surface of the Earth [Schuler 1923].

$$\frac{d\omega_1}{dt} = \frac{d\omega}{dt}. \tag{25}$$

Now

$$\frac{d\omega_1}{dt} = \frac{\gamma m a}{\Theta} \quad \text{and} \quad \frac{d\omega}{dt} = \frac{\gamma}{R},$$

where γ is the acceleration of the vehicle and R is the radius of the Earth. Thus equation (25) will be satisfied if

$$\gamma \cdot \frac{ma}{\Theta} = \gamma \cdot \frac{1}{R}.$$

This condition is always fulfilled for the case $\gamma = 0$ in which the vehicle moves with uniform velocity in a specified plane. If the vehicle accelerates, then the condition is not fulfilled unless

$$\frac{\Theta}{mga} = \frac{R}{g}.$$

The period of the pendulum is given in this case by equation (24) as

$$T = 2\pi\sqrt{\frac{R}{g}} = 84 \text{ minutes}.$$

This is the period that is obtained if one substitutes $l = R$ into the usual formula for the period of a simple pendulum with length l.

Schuler shows how a pendulum with the required period can be realized by a top with a vertical axis (a "gyropendulum" or an "artificial horizon") and by a top with a horizontal axis (a "gryocompass").

Schuler also postulates (but does not attempt to prove) the general theorem that "an oscillatory mechanical system, on whose center of gravity acts a central force, is not set into oscillation by arbitrary motions on a spherical shell about the center of force if its period of oscillation is equal to that of a simple pendulum with the length of the radius of the sphere in the acting field of force." He ends his paper, about which much has been written, with the following reflection:

> It is perhaps characteristic of human creation that I was led to this simple and clear relation directly by calculations with the complicated and obscure equations of the gyroscopic compass.

317. (page 863) Anschütz-Kaempfe addressed the cost of his gyroscopic compass in his closing remarks after the discussion of his presentation to the *Schiffbautechnische Gesellschaft*:

> I must still answer a question from the table of the Board of Directors regarding the price of the apparatus. The apparatus naturally consists of a complete installation, from the machine for its actuation, resistors, rotation counters, measuring instruments, etc., to the apparatus itself and its cable. Such an installation presently costs 20 000 M. (Agitation.) It is possible that the price may be reduced with factory production. In any case, the apparatus represents years of continuous work with immense costs, so that the price of the apparatus is thus explained.

318. (page 863) Gilbert Thomas Walker (1868–1958) was a mathematician in Cambridge University when he wrote his article for the *Encyklopädie der mathematischen Wissenschaften.* He later become director-general of the British observatories in India. Walker's article includes sections on billiards, golf balls, bicycles, and boomerangs [Walker 1904].

319. (page 864) Emmanuel Carvallo (1856–1954) was *directeur d'études* at the *École Polytechnique.* The first part of his paper on the dynamics of the bicycle considers the rolling motion and stability of a hoop (*cerceau*), the motion of a hoop struck by a stick (*baguette*), and the motion of a unicycle (*monocycle*) [Carvallo 1900]. The rider of the unicycle is assumed to be a rigid body whose center of gravity always lies "on the line of greatest slope that passes through the center of the wheel" (the line formed by the intersection of the plane of the wheel and the vertical plane through the axis of the wheel). This positioning is assumed to be accomplished "by the skill of the rider."

320. (page 864) William John Macquorn Rankine (1820–1872) was professor of civil engineering and mechanics in the University of Glasgow. His strangely cited paper in *The Engineer* is titled "On the dynamical principles of the motion of velocipedes" [Rankine 1869]. Rankine does not speak of creating a centrifugal force that uprights the bicycle, but reasons more rationally as follows. If the bicycle moves with constant speed in a straight line, it will not tilt if its center of mass is directly above the track of the wheels. If the bicycle turns to the right with constant speed, its inclination can remain constant if it leans at the proper angle to the right, so that the center of mass is to the right of the track of the wheels. Suppose now that the bicycle tilts to the right while it moves with constant speed in a straight line, so that the center of mass moves to the right of the track of the wheels. The rider should turn to the right in order to reach a state of motion in which the rightward tilt can remain constant, so that the bicycle will not tilt further and fall.

321. (page 865) Francis John Welsh Whipple (1876–1943) studied mathematics at Trinity College, Cambridge. His paper on the stability of the bicycle [Whipple 1899] was judged "worthy of honourable mention" in the competition for the 1899 Smith Prize. Whipple worked for most of his career in meteorology. From 1925 to 1939 he was supervisor of the Kew Observatory near London, where his father, George Mathews Whipple, had been supervisor from 1878 to 1893.

322. (page 865) In the second part of his paper in the *Journal de l'École Polytechnique*, Carvallo derives a modification to the Lagrange equations that can be used for systems with nonholonomic constraints [Carvallo 1901]. The same extension is described in a more recent book by the American astronomer William Duncan MacMillan (1871–1948), and applied by him to the nonholonomic problem of a sphere rolling on a plane [MacMillan 1960, pp. 332–341].

323. (page 865) "Lagrange equations of the first kind" usually refers to equations of motion written in terms of Cartesian coordinates, with Lagrange multipliers to account for holonomic or nonholonomic constraints. Whipple does not use such equations, but rather uses D'Alembert's principle in the form presented by Routh [Routh 1960, pp. 45–59]. "Lagrange equations of the second kind" refers to equations of motion written in terms of complete and independent generalized coordinates; these equations are valid only for holonomic systems.

324. (page 866) There is now a large body of literature on the stability of the bicycle. Recent reviews of the subject with links to extensive online supporting material are given by Jodi Kooijman, Jaap Meijaard, James Papadopoulos, Andrew Ruina, and Arend Schwab [Kooijman 2011; Meijaard 2007]. The claim that the gyroscopic effects of the wheels are necessary for stable upright rolling motion of a bicycle is generally incorrect, although it is correct for the particular bicycle considered here. The cited authors have shown that the claim is based in part on an algebraic error in the stability analysis of this section. The error and its implications are discussed in note 328 below.

325. (page 867) Charles Émile Ernest Bourlet (1866–1913) was professor of mathematics in the *Lycée Saint-Louis*, the *Conservatoire national des arts et métiers*, and the *École nationale supérieure des beaux-arts.* In 1892, Bourlet was asked to write a series of articles for the weekly journal *Le Cycle.* He "seized the occasion, with eagerness, to give form to the numerous reflections and observations that I had collected during four years of cycling" [Bourlet 1895]. His more theoretical studies were published, "not without a certain apprehension," as the *Traité des Bicycles et Bicyclettes.* Bourlet's "little volume had an unexpected success," and a *Nouveau Traité* in two volumes was published in 1898.

Bourlet was a member of the technical committee of the *Touring-Club de France* and an active proponent of Esperanto.

326. (page 867) The rolling of a cylinder on a plane is an example of a holonomic rolling motion [Landau 1960, p. 123].

327. (page 875) We have corrected equation (9) on p. 875. The factor $(h_2 \sin\sigma + r\cos\sigma)$ in the third term on the right-hand side is printed incorrectly as $h_2 \sin\sigma$ in the original edition.

328. (page 880) There are two important sign errors in the expression (13) on p. 880. The expression should read

$$
\begin{aligned}
& - gMh\cos\sigma(c_2 A_v + c_1 B_v)\frac{u}{l} \\
& + gB_{hv}\Big(\underset{\uparrow}{+}\, M_1 h_1 \sin\sigma + M_2 r \frac{c_1}{l}\cos\sigma\Big)u \\
(13) \quad & - gM_1h_1M_2h_2 l\sin\sigma\cdot u \underset{\uparrow}{+} gM_1M_2h_1c_1 r\cos\sigma\cdot u \\
& + \frac{c_2 + c_1}{l}\cos\sigma N[2Nu + Mhu^2],
\end{aligned}
$$

where the corrected signs are marked by arrows. The errors in this expression were discovered by Kooijman, Meijaard, Papadopoulos, Ruina, and Schwab as part of the reviews cited in note 324. With the corrected signs, there are three nongyroscopic positive terms in (13). For the particular bicycle parameters assumed by Whipple, these three terms are small, and the positive gyroscopic terms are indeed necessary in order obtain a stable motion. It does not follow from the corrected expression, however, that the nongyroscopic positive terms are generally small, and that the gyroscopic terms are generally necessary for stability. For a further discussion, we refer to the papers of the cited authors.

329. (page 881) We have corrected the expression (15) on p. 881. The factor $\cos\sigma$ before the final right parenthesis in the third term is incorrectly omitted in the original edition. This error has no consequence for the following discussion.

330. (page 884) Karl Gustaf Patrik de Laval (1845–1913) was an engineer, inventor, and member of the Swedish parliament. His contributions to the development of the steam turbine are described by Henry Winram Dickinson in his enjoyable (if Anglocentric) *Short History of the Steam Engine* [Dickinson 1939].

De Laval also invented a centrifugal cream separator that became widely used in the dairy industry.

331. (page 884) The encyclopedia article on technical thermodynamics was written by Moritz Schröter (1851–1925), director of the Institute for Theoretical Machine Research in Munich, and Ludwig Prandtl [Schröter 1906]. Section 20 discusses compressible flow in nozzles, and section 23 makes a brief mention of the steam turbines developed by Charles Algernon Parsons (1854–1931), de Laval, Auguste Camille Edmond Rateau (1863–1930), and Heinrich Zoelly (1862–1937).

332. (page 885) Föppl became interested in the dynamics of a rotor on a flexible shaft after Ludwig Klein (1868–1945), then an assistant at the *Technische Hochschule* in Munich, gave a lecture on the steam turbine to the Bavarian chapter of the *Verein deutscher Ingenieure* [Klein 1895]. Klein described de Laval's use of a flexible shaft, and compared a rotor on such a shaft to "a freely rotating body—a thrown stone, a spinning top, a heavenly body—which always aligns itself with the principal axis of free rotation through the center of gravity, so that the centrifugal forces cancel." Klein's lecture was followed by a lively discussion in which some "very competent and experienced engineers" questioned the self-centering of the rotor [Föppl 1895]. They argued that an eccentrically mounted rotor on an initially straight shaft experiences an unbalanced centrifugal force when the shaft begins to rotate. Since the force acts at the center of mass of the rotor and is directed away from the axis of rotation, the shaft must bend, and bend in such a way that the eccentricity is increased. The whirling must therefore be strengthened by the flexibility of the shaft.

When Klein brought this objection to Föppl, Föppl could not "escape the weight" of this "perfectly correct conclusion." Klein then "decided to appeal to the highest authority in these matters, the experiment." His experiments showed the "very beautiful" self-centering of a rotor on a flexible shaft, but also showed that this self-centering occurred only when the angular velocity exceeded a certain critical value. When the angular velocity of the rapidly rotating shaft was gradually reduced to a certain value, a strong whirling took place.

After Klein's experiment, Föppl considered it an "obligation of honor [*Ehrenpflicht*] to come to an understanding of this affair." He realized that the original objection is indeed correct, but that the initial assumption of an eccentric rotor on a straight shaft does not apply after the critical angular velocity is exceeded. After "long and arduous work," he succeeded to the extent that he could "represent the process theoretically, at least in its main features." Föppl's theory became the basis for all subsequent analyses of whirling phenomena.

333. (page 885) Aurel Boleslav Stodola (1859–1942) was a Slovak engineer who became professor of mechanical engineering at the *Polytechnikum* (now the *Eidgenössische Technische Hochschule*) in Zurich. His book on the steam turbine was translated into English by Louis Centennial Loewenstein (b. 1876) of Lehigh University [Stodola 1905].

334. (page 901) The Elberfeld–Barmen suspension railway continues to operate today as part of the *Wuppertaler Schwebebahn*, carrying 82,000 daily commuters. The railway system was designed by Eugen Langen (1833–1895), who had earlier collaborated with Nikolaus August Otto (1832–1891) in the founding of the first company to manufacture and sell gasoline engines.

335. (page 902) The Manchester–Liverpool monorail was proposed by Fritz Bernhard Behr (1842–1927), a naturalized British subject who was born in Berlin and educated in Paris. Behr adopted the rail system of the French engineer Charles François Marie-Thérèse Lartigue (1834–1907), who developed his mule-powered elevated monorail to avoid blockage by blown sand in Algeria. A cross section of Behr's monorail track and carriage* at the 1897 International Exhibition in Brussels is shown in Fig. 216. The height of the central running rail is 1.25 m, and there are two horizontally facing guide rails on each side of the A-frame trestle. The electrically powered carriage attained a maximum speed of 83 miles per hour and an average speed of 70 miles per hours over 5 kilometers, which Behr considered, under the circumstances, a success: "The examination by members of the Royal Commission, instituted by the Government of Belgium, clearly brought out the fact, that though the full speed proposed in the curves of 500 metres radius was not attained, these curves were so sharp, the gradients so steep, the power so small, and the carriage so badly constructed, that it was a much greater feat to attain even 70 miles per hour on such a line and on the curves, than to attain 110 miles on a properly constructed monorail, under such conditions as would arise in ordinary railway practice" [Behr 1902, p. 83].

Behr's design for the Manchester–Liverpool line proposed single-carriage trains ("couplings are a source of a danger, which could on no account be allowed") running at an average speed of 110 miles per

* "Internally the carriage is divided into compartments, one of which, 20 ft. long by 5 ft. 6 in. wide, being reserved for royalty. No expense has been spared on the luxurious fittings of the compartments, which are upholstered in Utrecht velvet, the joiner's work being in Spanish mahogany, and the roof linings of imitation leather." [*Engineering*, Vol. 63, 1897, p. 849.]

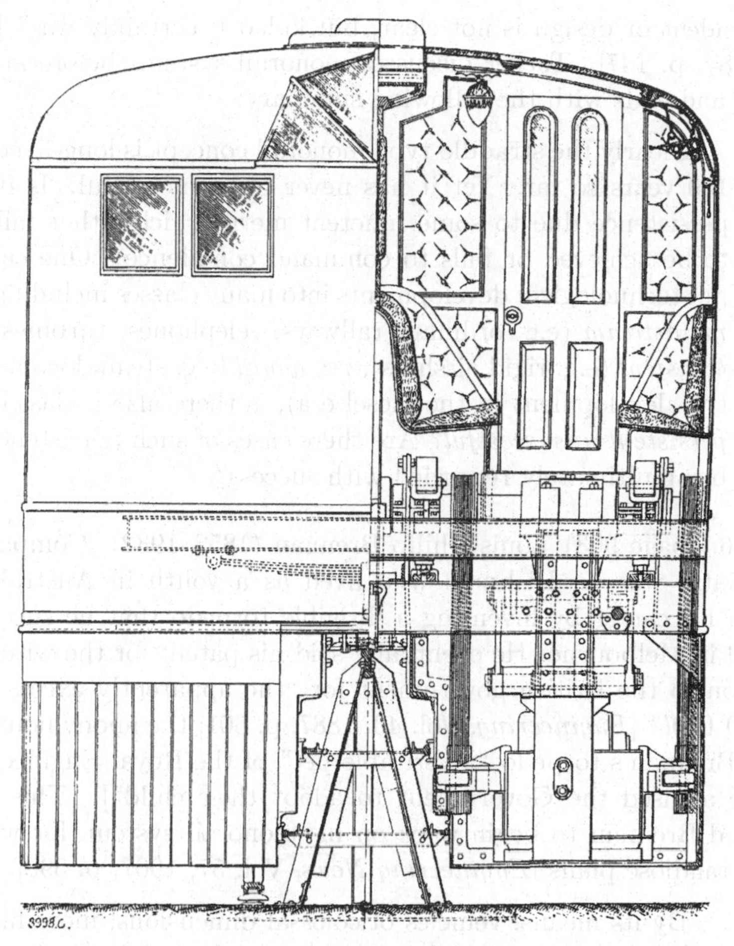

Fig. 216. Cross section of Behr's monorail carriage at the 1897 International Exhibition in Brussels [*Engineering*, Vol. 63, 1897, p. 854].

hour, making the $34\frac{1}{2}$ mile trip in 20 minutes (the shortest travel time by rail today is 47 minutes). His plan was approved in an Act of Parliament in 1901, but the act contained a clause giving the Board of Trade (B.O.T.) power to require further experiments, and this clause eventually doomed the project: "When in July 1903 a prospectus was issued inviting applications for a capital of £2.1 million, so few applications for shares were received that no issue was made. Evidently investors were not interested in a project which required a third to half a million pounds to be spent before knowing whether the B.O.T. would give their approval. Although the promoters and Behr did not give up, this was effectively the end. Whether the B.O.T. had killed the scheme by accident or design is not clear, but killed it certainly was" [Tucker 1983–84, p. 147]. Tucker discusses monorail systems before and after Behr, and ends with the following summary:

> Clearly the straddle-type monorail concept is long-lived; 150 years so far. Yet it has never been successful. Is its persistence due to some inherent merit which either fails to be achieved or fails to command confidence? One can put technological developments into many classes including: *mainstream* (e.g. ordinary railways, telephones, turbines), *transient* (e.g. rigid airships), *rearguard* (e.g. steam locomotive development in the diesel era); is there also a class of *persistent unsuccessful?* Are there cases of such persistence being ultimately rewarded with success?

336. (page 903) Louis Philip Brennan (1852–1932), Companion of the Bath, was an Irishman who lived as a youth in Australia. He began his career by inventing a dirigible torpedo that he successfully tested in Melbourne. He eventually sold his patent for the wire-guided weapon to the British government for "the apparently excessive sum of 110,000*l*" [*Engineering*, Vol. 43, 1887, p. 601; the anonymous article calls Brennan's torpedo "a veritable pet" of the Royal Engineers, who "have advised the Government to adopt their child"]. This success allowed Brennan to begin work on his monorail system, for which he had grandiose plans [*Engineering News*, Vol. 57, 1907, p. 598]:

> By its means, vehicles of colossal dimensions, more like moving hotels or ships on shore than the usual form of railway carriage, will travel at velocities of probably from two to three times that attainable at present; at the same time there will be a complete absence of lateral oscillation in the carriages, and a great increase in the smoothness with which they will run * * * every facility can be given for the enjoyment of life while travelling, as the carriage may be fitted up with all the comforts of a first-class hotel, and separate rooms provided for every purpose, such as smoking, dining, reading, writing, etc., even promenade and music rooms will probably be provided to relieve the tedium of railway travelling.

A small (length 6 ft., weight 300 lb) working model of Brennan's gyroscopically stabilized monorail carriage was exhibited at the May 1907

Soireé of the Royal Society, which also featured a water tank model of Schlick's gyroscopic ship stabilizer [*Engineering*, Vol. 83, 1907, p. 623]. In November 1909, Brennan gave a well-received demonstration of a larger carriage (length 40 ft., weight 22 tons with a load capacity of 10–15 tons), but could not raise funds for further development, and his monorail project was abandoned in 1912 [Tomlinson 1980]. Brennan then devoted thirteen years to the design of a helicopter that crashed during a test flight in 1925. Undeterred, he began work at the age of 74 on a privately funded gyroscopically stabilized two-wheel automobile. A full-sized working prototype was presented to the British automotive firms Morris, Austin, and Rover, all of which declined to produce the vehicle. The prototype was donated to the museum of the Sperry Gyroscope Company and was destroyed by a bomb in World War II. But the technical and popular fascination with the gyroscope was illustrated yet again at the 1961 International Automobile Show in New York, where Ford Motor Company exhibited a (nonfunctional) display model of the Gyron, which was "designed to run on two wheels (motorcycle fashion) and relies on a gyroscope to keep it upright" [*Car Life*, May 1961, p. 9]. George W. Walker, Ford vice president and director of styling, added that "its present cost alone makes it impractical for production, either now or in the forseeable future."

Fig. 217. Display model of the Ford Gyron [Huntington 1961, p. 43].

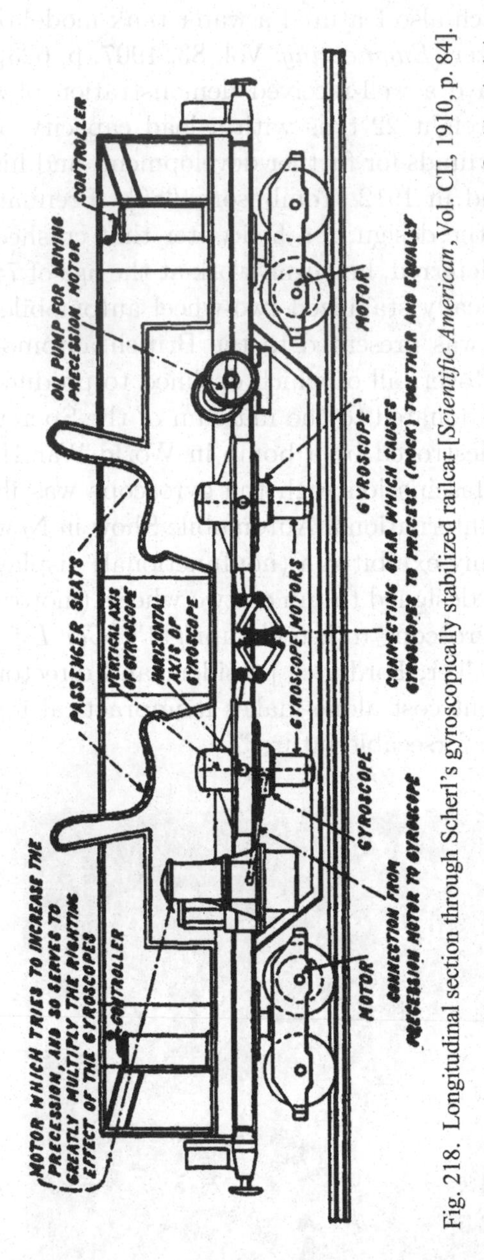

Fig. 218. Longitudinal section through Scherl's gyroscopically stabilized railcar [*Scientific American*, Vol. CII, 1910, p. 84].

337. (page 903) August Hugo Friedrich Scherl (1849–1921) was a German newspaper and magazine publisher. A 1921 obituary in the *New York Times* calls him the "pet publisher of the Kaiser," and credits (blames?) him for introducing "what passes here [Berlin] as the American style of journalism." The obituary also mentions Scherl's other pursuits: "Some ten years ago Scherl interested himself in a 'monorail' system and other public and industrial enterprises which were less successful and reduced his fortune considerably." Scherl's gyroscopically stabilized single-rail car (Fig. 218) was brought to the United States and exhibited in Brooklyn in 1910. The vehicle was designed and built at the *Technische Hochschule* in Dresden under the direction of Scherl's son Richard [Kübler 1909]. The significance of the motor that drives the precession of the gyroscopic rotors is discussed on pp. 908–911. In 1909, Scherl published *Ein Neues Schnellbahnsystem*, an attractive little book in which he proposed his monorail system as part of a comprehensive solution to the "crisis in contemporary railway affairs" [Scherl 1909]. Scherl designed a network of rail lines for all of Germany, with utopian drawings of elevated tracks and stations in large cities. Such a system, Scherl argued, could be realized only by electrically powered single-rail trains operating at speeds of 200–250 km/hr.

338. (page 904) Föppl's conceptual model of a single-rail carriage is shown in Fig. 219. The carriage is represented by a pendulum that carries two large masses C and D. The masses can be fixed at various locations along the rod BB, so that the center of mass of the pendulum can be above, below, or at the pivot A. The three principal positions for the rotor of the gyroscopic stabilizer are labeled *I*, *II*, and *III*.

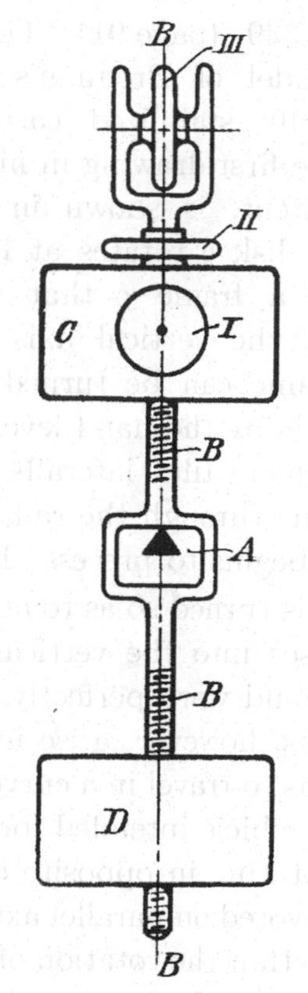

Fig. 219. Conceptual representation of a gyroscopically stabilized single-rail carriage [Föppl 1906].

If the axes of the rotors are fixed with respect to the pendulum and the angular velocities of the rotors about their axes are constant, then the motion of the pendulum is completely unaffected by the spin of the rotors. "One may be convinced of this by experiment," writes Föppl, "and it is also easy to give a theoretical proof. But whoever wishes to concern himself with such matters without previous detailed knowledge of the theory of the top may not expect to obtain everything he lacks from a short article." Either the angular velocities of the rotors or the orientation of the rotor axes with respect to the pendulum must be variable if the rotors are to influence the motion of the pendulum. For position *I*, the desired influence can be obtained with a fixed axis by making the angular acceleration of the rotor depend on the angular displacement of the pendulum. (Föppl describes a hypothetical system in which a human conductor senses the tilt of a monorail carriage and operates a lever that controls the current to an electric motor that drives a rotor in position *I*. Föppl compares such a conductor to the rider who stabilizes a bicycle through appropriate manipulation of the handlebars and displacements of his center of gravity: "Presumably it would require even less skill, and be easier to learn to steer the monorail carriage than the bicycle.") For positions *II* or *III*, the desired influence can be obtained by mounting the axis of the rotor in a frame that can rotate with respect to the pendulum, as in the ship stabilizer or Brennan's monorail carriage.

339. (page 911) The simplest model of Brennan's gyroscopically stabilized carriage, and the first drawing in his 1905 US patent, is shown in Fig. 221. A disk b rotates at high speed in a frame c that is pivoted on the vertical axis de. The frame can be turned about its axis by the hand lever h. If the vehicle tilts laterally about an axis through the rail, the frame c begins to precess. If the lever h is turned so as to accelerate the precession, "then the vehicle at once rises into the vertical position again. A vehicle as above described would work perfectly, for example, on a single straight track. Difficulties, however, arise in the use of a single gyrostat whenever a vehicle has to travel in a curved path and be reversed. I therefore in the case of a vehicle intended for ordinary purposes employ two similar gyrostats rotating in opposite directions. The carriers of the two gyrostats are pivoted on parallel axes, and I connect the carriers by means of gearing, so that the rotation of one carrier in one direction insures a corresponding rotation of the other carrier in the opposite direction. I then control the movements of the carriers as before by means of a suitable lever." [Brennan 1905].

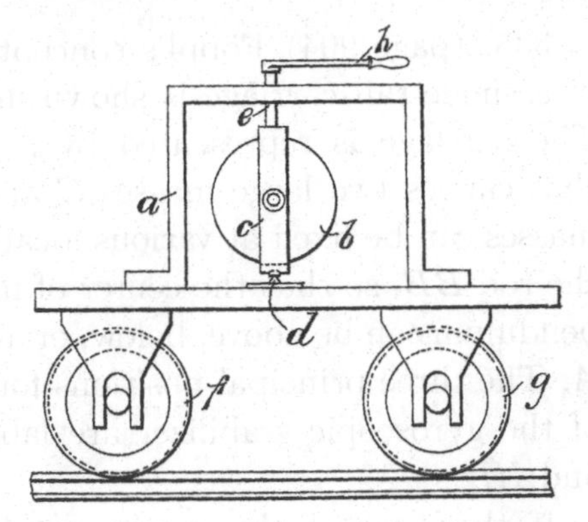

Fig. 221. Gyroscopically stabilized single-rail carriage [Brennan 1905].

The automatic rolling mechanism for accelerating the precession of the rotors in Brennan's carriage is shown in Fig. 222. The counterrotating rotors F and F' are mounted in the evacuated casings G and G', and are driven by electric motors not shown in the drawing. The casings can rotate about the vertical axes EE and $E'E'$. The casings are coupled by the gears J and J', so that when the axle H rotates into the plane of the drawing, so does the axle H'. The entire rotor assembly is pivoted about a horizontal axis through C.

If the carriage tilts for some reason to the right, the shelf D will come into contact with the spinning axle H. The axle H will then roll into the plane of the drawing, and the resulting precession of the two rotors will produce a moment that tilts the carriage back to the left. The carriage will rotate too far to the left, and the loose roller M' will thus come into contact with the shelf L', causing a precession that

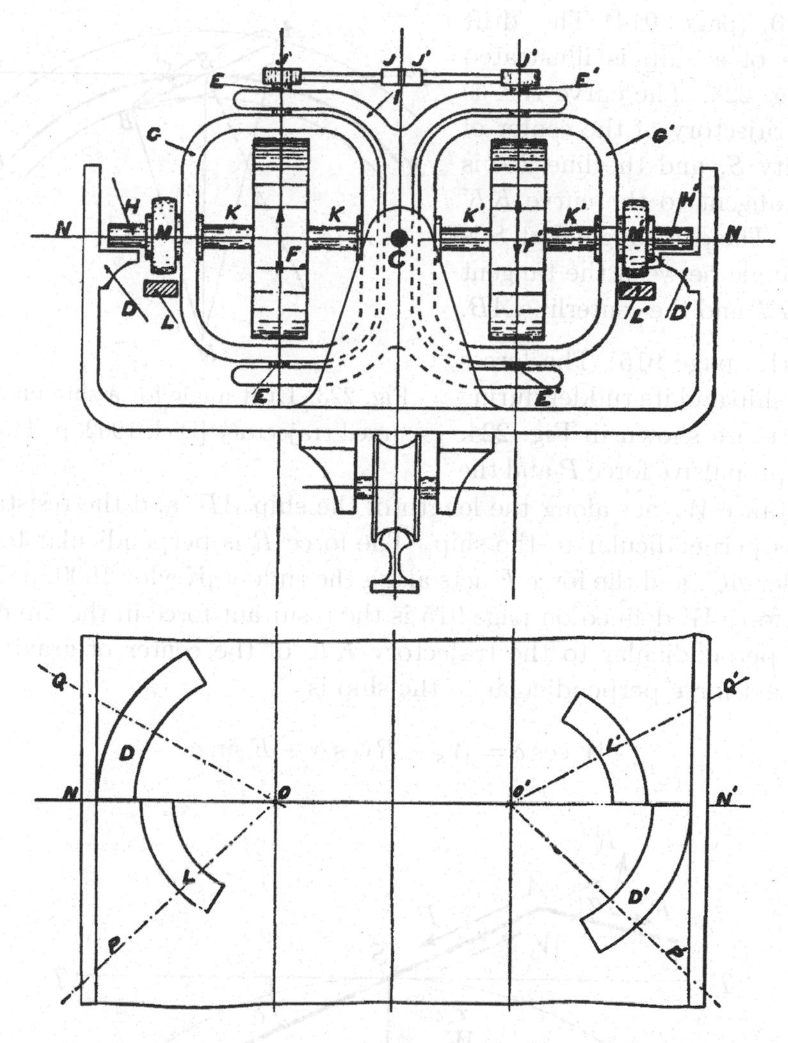

Fig. 222 Rolling mechanism for accelerating the precession of the rotors in Brennan's monorail carriage [Perry 1957, pp. 91–92].

tilts the carriage back to the right and restores the rotors to their original positions.

According to John Perry, Brennan considered the rolling mechanism in Fig. 222 to be a crude method of accelerating the precession [Perry 1957, p. 93]. A description of a more sophisticated system that was implemented by Brennan is described by Harold Crabtree in his *Spinning Tops and Gyroscopic Motion* [Crabtree 1913, p. 95].

340. (page 914) The drift angle of a ship is illustrated in Fig. 223. The curve KK is the trajectory of the center of gravity S, and the line TT is the tangent to the curve KK at S. The drift angle δ at S is the angle between the tangent line TT and the centerline AB.

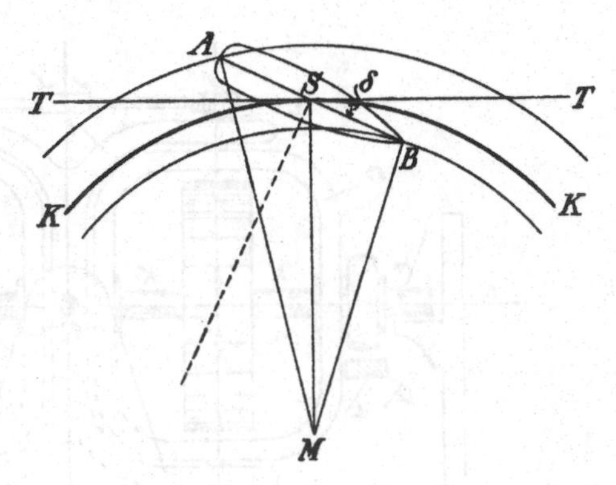

Fig. 223. Drift angle for a ship on a curved trajectory [Dick 1902, p. 445].

341. (page 915) The forces on a ship and its rudder during a turn are shown in Fig. 224. The propulsive force P and the resistance W_e act along the length of the ship AB, and the resistance W_s is perpendicular to the ship. The force R is perpendicular to the rudder AC, and the force F acts along the rudder [Krylov 1900, p. 558]. The force W defined on page 915 is the resultant force in the direction SM perpendicular to the trajectory KK of the center of gravity S. The net force perpendicular to the ship is

$$W \cos \delta = W_s - R \cos \alpha + F \sin \alpha.$$

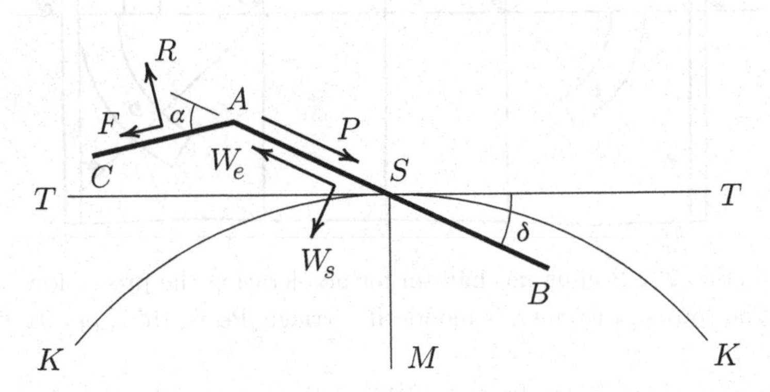

Fig. 224. Forces on a ship during a turn.

342. (page 917) A brief article on the loss of the British torpedo-boat destroyer Cobra appeared in the September 27, 1901, issue of *Engineering* (Vol. 72, pp. 452–453.) The Cobra and her sister ship Viper were the first two torpedo-boat destroyers fitted with Parsons steam turbines. The Cobra broke in two and sank on her maiden voyage from

the Tyne to Portsmouth; the Viper was stranded a few weeks before. According to the article,

> The Cobra was built by Messrs. Armstrong, Whitworth, and Co. at Elswick, and engined by the Parsons Marine Steam Turbine Company at Wellsend-on-Tyne. She left the Tyne for the purpose of being delivered at Portsmouth on Tuesday, September 17, in charge of Lieutenant Alan Bosworth Smith, who had under his command a navigating party of 54 in all. There were also 25 other persons on board, including Mr. Magnus Sandison, the chief engineer to Elswick Shipyard, and Mr. Robert Barnard, the manager of the Parsons Marine Steam Turbine Company. Both of the gentlemen were lost; as was also Lieutenant Bosworth Smith, who fully maintained the reputation of British seamen by remaining at his post to the last.

The cause of the wreck could not be immediately determined. There were reports that the ship had struck a shoal, but the depth of the water in the place where the ship is said to have struck was six or seven fathoms, and she drew only six or seven feet. A week after the article, a letter to the editor of *Engineering* raised the possibility that the gyroscopic effects of the turbine might have contributed to the loss of the ship:

> SIR,—The startling announcement has appeared in last week's daily press that the ships sent to explore the site of the Cobra's wreck have failed to discover any trace of a rock upon which she could have struck. Some other cause must therefore be sought for her breaking asunder amidships so suddenly and without warning. May the suggestion be offered whether the explanation may possibly be found in the superadded gyroscopic action of the twelve turbines themselves, rotating at so high a speed as 1200 revolutions per minute in planes transverse to the ship's length? Being unable at present to refer to the full details given in your vol. lxviii. (pages 191, 221, and 256), I have recourse for the moment to the statement in vol. lxix. (page 219) respecting the turbines in the sister-ship, the Viper. Here the weight of the propellers, shafting, &c, is given as $7\frac{3}{4}$ tons, to which apparently must be added the weight of the twelve turbines. The photograph of the Viper in page 219 shows her steam-

> ing in a smooth sea, where gyroscopic action would not come into play. But in a rough sea, each wave she breasts would throw a bending strain upon her amidships, alternately upwards and downwards; as in testing an axle under a falling weight, and reversing it between each blow. To such alternate straining, aggravated by the gyroscopic resistance, must the Cobra have been subjected by the heavy seas she encountered from the moment of quitting the Tyne till she so suddenly broke in two.
>
> Would any of your many competent readers think it worth while to investigate this suggestion?—such as the Hon. Charles A. Parsons himself, Mr. R. Edmund Froude, Mr. J. Macfarlane Gray, Mr. A. Malloch, Mr. Beauchamp Tower, Mr. C. Humphrey Wingfield, Mr. Sydney Barnaby, or any of the college professors of engineering, whose mathematics have not rusted like those of
>
> IGNORAMUS.

The interest in the theory of the gyroscope at the time is shown by the series of contentious (and often amusing) letters that followed.* Macfarlane Gray replied twice, and did not miss the chance to recall his previous exposé of Bessemer's error:

> As, however, many of your readers are keenly desirous to have this, their first mechanical paradox, explained to them—in England, "Why does a top stand whipping when spinning?" in Scotland, "Hoo diz a peerie staund up whun it's soondin?" and as the lack of this knowledge cost a London syndicate, principally engineers, a quarter of a million

*The archives of *Engineering* are a seemingly limitless source of inspiration and enjoyment. Opening to a random page in a large volume, one may find, for example, a report on a lecture by John Perry to the British Association, a lecture in which Perry urged the criterion of usefulness for establishing a curriculum in mathematics. "But by usefulness he does not understand what one might at once think of. He set up a long table of obvious forms of usefulness which would accrue from the study of mathematics, comprising higher emotions, mental pleasures, and brain development; passing examinations—the only form hitherto recognised by teachers and not neglected—giving a man mental tools as easy to use as his arms and legs; teaching him the importance of thinking for himself; making him feel in any profession that he knows the principles on which it was founded; giving acute philosophical minds counsel of perfection and preventing the attempt to develop any philosophical subject from the purely abstract point of view."

> sterling in 1874, and it is probable that those of them who are still on this side—I cannot speak for the others—may not even yet have come into possession of the knowledge for which they paid so much, I will in this letter give a somewhat elaborate explanation of how the gyroscopic couple is set up, and its magnitude.

Official investigations into the cause of the loss were contradictory. A Court Martial in 1901 concluded that "HMS Cobra did not touch the ground nor meet with any obstructions nor was her loss due to any error in navigation, but is attributable to the structural weakness of the ship. The Court also finds that Cobra was weaker than any other destroyers and in view of that fact it is to be regretted that she was purchased into His Majesty's service" [Faulkner 1985]. A special Committee on Torpedo-Boat Destroyers, however, was "unable to attribute the breaking of Cobra to the action of the sea alone, supposing the material of her structure to have been free from flaws and properly connected." A reassessment of available data in the cited paper by Faulkner concludes that the failure was most likely due to "the quite inadequate compression strength of the hull." No mention of gyroscopic effects is made in the reassessment.

343. (page 918) The first-class passenger and freight steamer Creole was built by the Fore River Shipbuilding Company in Quincy, Massachusetts. She was powered by two 4000 hp steam turbines designed by the American engineer Charles Gordon Curtis (1860–1953). Detailed information on the design and performance of the turbines is included in the cited article in *Engineering* [Edwards 1907]. The Creole was put into service between New York and New Orleans, but was temporarily withdrawn after her first round trip. Mud from the Mississippi River was drawn into the boilers and eventually made its way into the air-pumps, resulting in their disablement on the return journey to New York.

344. (page 919) See note 220 in our translation of Vol. III of the *Theorie des Kreisels.*

345. (page 919) Georges Ernest Fleuriais (1840–1895) was a French naval officer who invented navigational instruments, made astronomical and geographic observations in the Pacific, and participated in the repression of the Paris Commune. A sextant with Fleuriais's gyroscopic

horizon is shown in Fig. 225. The support point of the top in the cylindrical chamber is marked P, and the two plano-convex mirrors on the upper surface of the top are marked V. A later version of Fleuriais's gyroscopic horizon with a slightly different optical system is described and illustrated in *Engineering*, Vol. 79, 1905, pp. 361–362.

346. (page 921) The cited article by Ludwig Prandtl is the first paper in the first volume of the *Zeitschrift für Flugtechnik und Motorluftschiffahrt.* Prandtl was the scientific editor of the journal, and made some remarks on his view of his role [Prandtl 1910]:

> In order to indicate my intentions more closely, I may emphasize in advance that I will not accept the common claptrap about the contrast between theory and practice. The contrast lies, in my perception, between good and bad, correctly and incorrectly applied theories; a good theory is in accordance with the results of practical experience, or at least reproduces the essential character of the experimental facts. A theory that does not achieve this is a poor theory, and does not deserve to be considered further. If we ask for the causes that can lead to the failure of a theoretical consideration, we find, in the vast majority of cases, that it has begun from inapplicable assumptions, or that a theory, valid in itself under certain assumptions, has been applied to cases that are incompatible with those assumptions; much more seldom, that an error has been made in the logical sequence of conclusions that is built on the assumptions. It is clear that a similar statement can be made about experiments; here too the possibility of drawing false conclusions by the nonconsideration of important circumstances is present to a high degree.
>
> Theoretical articles and reports on experiments will be gladly accepted if they can be considered as good theories and good experiments in the previously indicated sense; on the other hand, I count it among my duties to protect the reader of this journal from false or merely useless theories and the like, holding myself by no means infallible in this. So much by way of introduction.
>
> In the following presentation of some relations from mechanics that are of interest in aeronautics, I also intend to point out certain frequently unclear or incorrect conceptions, thus hoping to exclude them, as much as possible, from this new journal.

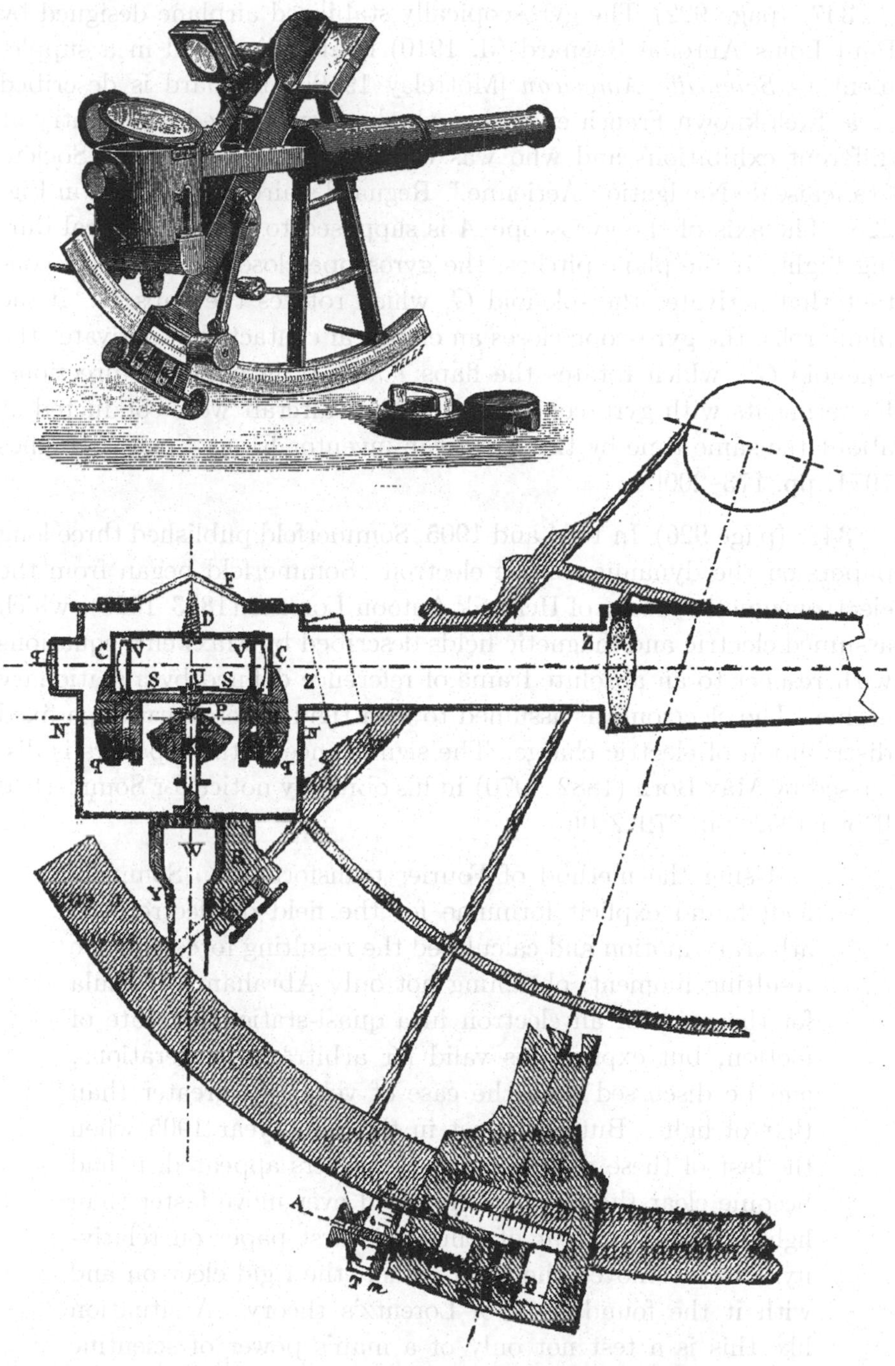

Fig. 225. Sextant with Fleuriais's gyroscopic horizon [Bud 1998, p. 32; Fleuriais 1886, p. 503].

347. (page 922) The gyroscopically stabilized airplane designed by Paul Louis Antoine Regnard (d. 1910) is also discussed in a supplement to *Scientific American* [Mottelay 1910]. Regnard is described as a "well-known French engineer, who has represented his country at different exhibitions and who was the first president* of the Société Française de Navigation Aérienne." Regnard's aircraft is shown in Fig. 226. The axis of the gyroscope A is supposed to remain vertical during flight. If the plane pitches, the gyroscope closes an electrical contact that activates the solenoid G, which rotates the flaps H. If the plane rolls, the gyroscope closes an electrical contact that activates the solenoid G^1, which rotates the flaps I and I^1 in opposite directions. Experiments with gyroscopically stabilized aircraft were conducted at about the same time by the American inventor Elmer Sperry [Hughes 1971, pp. 173–200].

348. (page 926) In 1904 and 1905, Sommerfeld published three long papers on the dynamics of the electron. Sommerfeld began from the electromagnetic theory of Hendrik Antoon Lorentz (1853–1928), which assumed electric and magnetic fields described by Maxwell's equations with respect to an absolute frame of reference defined by a stationary aether. The electron was assumed to be a rigid sphere carrying a fixed distribution of electric charge. The significance of these papers is discussed by Max Born (1882–1970) in his obituary notice for Sommerfeld [Born 1952, pp. 279–280]:

> Using the method of Fourier transforms he [Sommerfeld] found explicit formulae for the field of electrons in arbitrary motion and calculated the resulting force and the resulting moment, obtaining not only Abraham's formula for the mass of an electron in a quasi-stationary state of motion, but expressions valid for arbitrary accelerations, and he discussed even the case of velocities greater than that of light. But alas, just in the same year 1905 when the last of these three formidable papers appeared, it had become clear that no particle could ever move faster than light: it was the year of Einstein's first paper on relativity, which removed the fixed aether, the rigid electron and with it the foundations of Lorentz's theory. A situation like this is a test not only of a man's power of scientific

*He was president, but not the first; see the journal *L'Aéronaute, Bulletin mensuel international de la navigation aérienne*, http://gallica.bnf.fr.

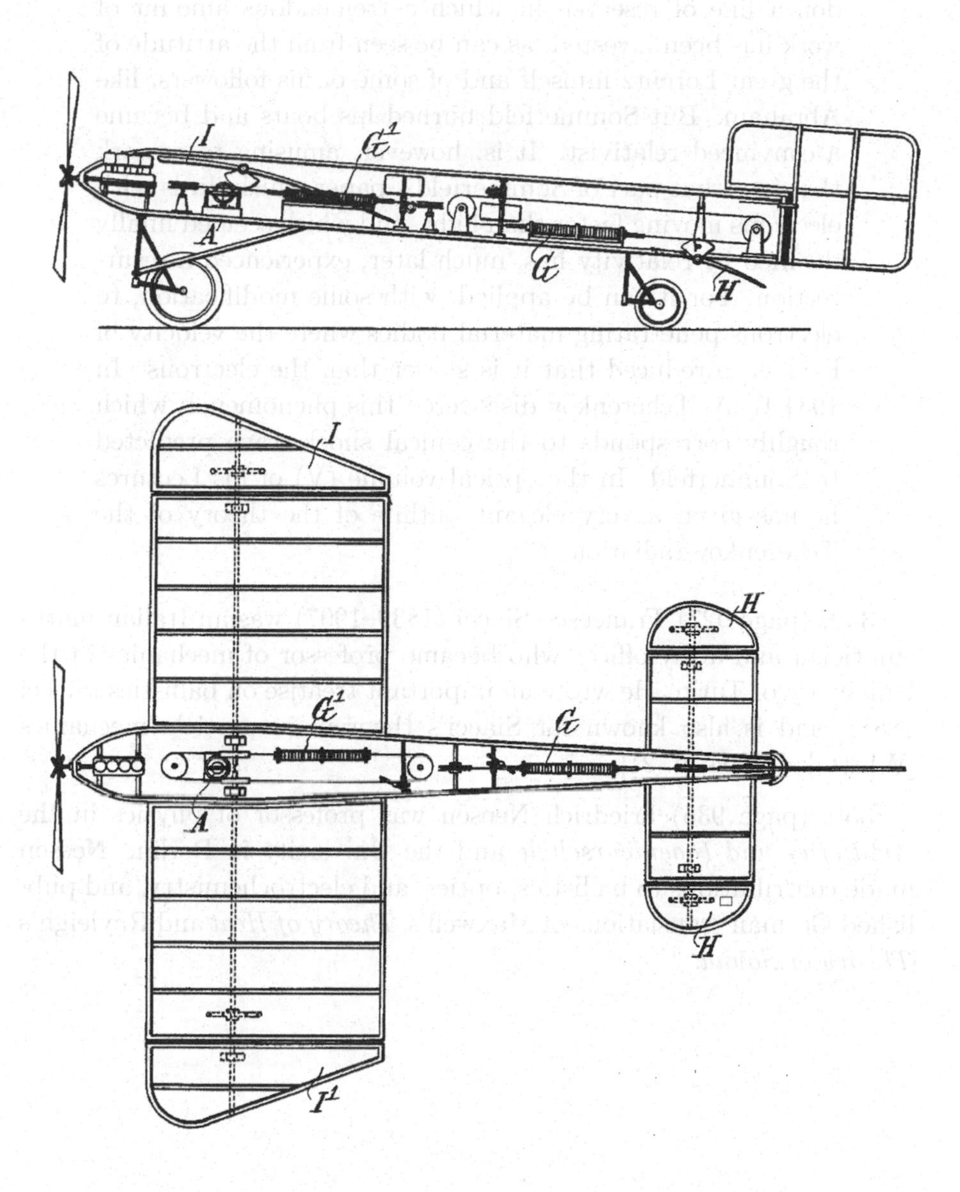

Fig. 226. Regnard's aircraft with gyroscopically activated control surfaces [Regnard 1912].

judgment but also of his character. It is not easy to abandon a line of research in which a tremendous amount of work has been invested, as can be seen from the attitude of the great Lorentz himself and of some of his followers, like Abraham. But Sommerfeld burned his boats and became a convinced relativist. It is, however, amusing to remark that just that part of Sommerfeld's paper which deals with electrons moving faster than light, and which seemed finally doomed by relativity has, much later, experienced a resurrection. For it can be applied, with some modification, to electrons penetrating material bodies where the velocity of light is so reduced that it is slower than the electrons. In 1934 R. A. Tcherenkov discovered this phenomenon which roughly corresponds to the conical shock wave predicted by Sommerfeld. In the optical volume (V) of his Lectures he has given a very elegant outline of the theory of the Tcherenkov radiation.

349. (page 927) Francesco Siacci (1839–1907) was an Italian mathematician and army officer who became professor of mechanics in the University of Turin. He wrote an important treatise on ballistics [Siacci 1888], and is also known for Siacci's theorem in particle mechanics [Whittaker 1964, p. 21].

350. (page 935) Friedrich Neesen was professor of physics in the *Artillerie- und Ingenieurschule* and the University in Berlin. Neesen made contributions to ballistics, optics, and electrochemistry, and published German translations of Maxwell's *Theory of Heat* and Rayleigh's *Theory of Sound.*

References.

[Anschütz-Kämpfe 1909]. Hermann Anschütz-Kämpfe, "Der Kreisel als Richtungsweiser auf der Erde mit besonderer Berücksichtigung seiner Verwendbarkeit auf Schiffen," *Jahrbuch der Schiffbautechnischen Gesellschaft*, Vol. 10, 1909, pp. 352–369.

[Behr 1902]. Fritz Bernhard Behr, "High-Speed Electrical Railways and the Proposed Monorail between Manchester and Liverpool," *Transactions of the Liverpool Engineering Society*, Vol. XXIII, 1902, pp. 76–97; http://books.google.com.

[Behrens 1911]. Wilhelm Behrens, "Ein der Theorie der Laval-Turbine entnommenes mechanisches Problem, behandelt mit Methoden der Himmelsmechanik," Inaugural Dissertation, Georg-August-Universität in Göttingen, 1911, http://hathitrust.org; also *Zeitschrift für Mathematik und Physik*, Bd. 59, 1911, pp. 337–390.

[Berger 1904]. Franz Berger, "Zur Theorie des Schlickschen Schiffskreisels," *Zeitschrift des Vereines deutscher Ingenieure*, Vol. 48, 1904, pp. 982–983.

[Bessemer 1905]. Henry Bessemer, *Sir Henry Bessemer, F. R. S.*, Offices of "Engineering," London, 1905; http://books.google.com.

[Born 1952]. Max Born, "Arnold Johannes Wilhelm Sommerfeld, 1868–1951," *Obituary Notices of Fellows of the Royal Society*, Vol. 8, No. 21, 1958, pp. 274–296; http://jstor.org.

[Bourlet 1895]. Charles Émile Ernest Bourlet, *Traité des Bicycles et Bicyclettes*, Gauthier-Villars, Paris, 1895; http://hathitrust.org.

[Brecht 1910]. Gustav Brecht, "Schwerpunktslage u. Kreiselwirkungen bei elektrischen Lokomotiven," *Elektrische Kraftbetriebe u. Bahnen*, Jahrgang VIII, 1910, pp. 121–127, 144–150, 161–167.

[Brennan 1905]. Louis Brennan, "Means for Imparting Stability to Unstable Bodies," US Patent 796,893, 1905.

[Broelmann 1998]. Jobst Broelmann, "The Development of the Gyrocompass—Inventors as Navigators," *Journal of Navigation*, Vol. 51, No. 2, 1998, pp. 267–276.

[Broelmann 2002]. Jobst Broelmann, *Intuition und Wissenschaft in der Kreiseltechnik, 1750 bis 1930*, Deutsches Museum, Munich, 2002.

[Bud 1998]. Robert Bud and Deborah Jean Warner, *Instruments of Science: An Historical Perspective*, Garland Publishing Co., New York, 1998.

[Bureau of Ordnance 1898]. *The Whitehead Torpedo*, Bureau of Ordnance, Department of the Navy, Naval Torpedo Station, Newport, R.I., 1998; http://www.hnsa.org.

[Carvallo 1900]. Emmanuel Carvallo, "Théorie du mouvement du monocycle et de la bicyclette. Première partie. Cerceau et monocycle," *Journal de l'École Polytechnique*, Sér. II, Vol. 5, 1900, pp. 119–188; http://books.google.com.

[Carvallo 1901]. Emmanuel Carvallo, "Théorie du mouvement du monocycle et de la bicyclette. Deuxième partie. Théorie de la bicyclette," *Journal de l'École Polytechnique*, Sér. II, Vol. 6, 1901, pp. 1–118; http://books.google.com.

[Crabtree 1909]. Harold Crabtree, *An Elementary Treatment of the Theory of Spinning Tops and Gyroscopic Motion*, Longmans, Green and Co., London, 1909; http://books.google.com.

[Cranz 1910]. Carl Cranz, *Lehrbuch der Ballistik*, B. G. Teubner, Leipzig, 1910; http://books.google.com.

[Cranz 1944]. Carl Cranz, *A translation of Cranz's Textbook of ballistics*, United States Office of Scientific Research and Development, Technical Reports Section for Armor and Ordnance, National Defense Committee, Washington, 1944.

[Dick 1902]. Carl Dick and Otto Kretschmer, *Handbuch der Seemannschaft*, Part 2, Ernst Siegfried Mittler und Sohn, Berlin, 1902; http://books.google.com.

[Dickinson 1939]. Henry Winram Dickinson, *A Short History of the Steam Engine*, University Press, Cambridge, 1939.

[Diegel 1899]. Carl Diegel, "Selbstthätige Steuerung der Torpedos durch den Geradlaufapparat," *Marine-Rundschau*, 10. Jahrgang, Heft 5, 1899, pp. 517–550; http://books.google.com.

[Dubois 1864]. Edmond Dubois, *Théorie du Mouvement des Corps Célestes*, Bertrand, Paris, 1864; http://gallica.bnf.fr.

[Dubois 1884]. Edmond Dubois, "Sur le gyroscope marin," *Comptes Rendus*, t. 98, 1884, pp. 227–229.

[Eckert 2013]. Michael Eckert, *Arnold Sommerfeld: Science, Life and Turbulent Times, 1868–1951*, English translation by Tom Artin, Springer, 2013.

[Edwards 1907]. Charles B. Edwards, "Builders' Trials of Curtis Turbine Steamer 'Creole,'" *Engineering*, Vol. 84, 1907, pp. 695–698; http://books.google.com.

[Faulkner 1985]. J. A. Faulkner, J. D. Clarke, and D. Faulkner, "The Loss of HMS COBRA—A Reassessment," *The Naval Architect*, May 1985, pp. 125–151.

[Fleuriais 1886]. Georges Ernest Fleuriais, "Gyroscope-Collimateur, Substitution d'un Repère Artificiel a l'Horizon de la Mer," *Revue Maritime et Coloniale*, Vol. 91, 1886, pp. 452–518; http://gallica.bnf.fr.

[Föppl 1895]. August Föppl, "Das Problem der Laval'schen Turbinenwelle," *Der Civilingenieur*, Bd. 16, 1895, pp. 333–342; http://hathitrust.-org.

[Föppl 1904a]. August Föppl, "Die Theorie des Schlickschen Schiffskreisels," *Zeitschrift des Vereines deutscher Ingenieure*, Vol. 48, 1904, pp. 487–483.

[Föppl 1904b]. August Föppl, "Zuschriften an die Redaktion," *Zeitschrift des Vereines deutscher Ingenieure*, Vol. 48, 1904, pp. 983–984.

[Föppl 1910]. August Föppl, *Vorlesungen über technische Mechanik*, Vol. VI, B. G. Teubner, Leipzig, 1910; http://books.google.com.

[Gerlach 1931]. Walther Gerlach and Arnold Sommerfeld, "Hermann Anschütz-Kaempfe," *Die Naturwissenschaften*, Vol. 19, 1931, pp. 666–669.

[Greenhill 1911]. George Greenhill, "Gyroscope and Gyrostat," *Encyclopædia Britannica*, Vol. XII, University Press, Cambridge, 1911, pp. 769–779, http://books.google.com.

[Huntington 1961]. Roger Huntington, "Are 4 wheels 2 too many?" *Car and Driver*, May 1961, pp. 43–45, 110–111.

[Hughes 1971]. Thomas Parke Hughes, *Elmer Sperry, Inventor and Engineer*, Johns Hopkins Press, Baltimore, 1971.

[k. u. k. Kriegsmarine 1893]. *Rangs- und Eintheilungs-liste der k. u. k. Kriegsmarine*, kaiserlich-königlichen Hof- und Staatsdruckerei, Wien, 1893; http://books.google.com.

[Klein 1907]. Felix Klein to Arnold Sommerfeld, 20 November 1907, Deutsches Museum Archive, HS 1977-28/A,170.

[Klein 1922]. Felix Klein, *Gesammelte Mathematische Abhandlungen*, Band 2, edited by R. Fricke and H. Vermeil, Springer, 1922; http://archive.org.

[Klein 1895]. Ludwig Klein, "Theorie, Konstruktion und Nutzeffekt der Dampfturbinen," *Zeitschrift des Vereines deutscher Ingenieure*, Bd. XXXIX, No. 40, 1895, pp. 754, 1189–1195.

[Kötter 1904]. Fritz Kötter, "Die Kreiselwirkung der Räderpaare bei regelmäßiger Bewegung des Wagens in kreisförmigen Bahnen," *Sitzungsberichte der Berliner Mathematischen Gesellschaft*, Feb. 1904, pp. 36–45; appendix to *Archiv der Mathematik und Physik*, Bd. 7, 1904; http://books.google.com.

[Kooijman 2011]. J. D. G. Kooijman, J. P. Meijaard, J. M. Papadopolous, A. Ruina, and A. L. Schwab, "A bicycle can be made self-stable without gyroscopic or caster effects," *Science*, Vol. 332, Issue 6027, 2011, pp. 339–342; http://www.sciencemagazine.org.

[Krylov 1901]. Alexei Nikolaevich Krylov and Conrad Heinrich Müller, "Die Theorie des Schiffes," *Encyklopädie der mathematischen Wissenschaften*, Band IV, Teil 3, Art. 22, B. G. Teubner, Leipzig, 1901–1908, pp. 517–593; http://gdz.sub.uni-goettingen.de.

[Kübler 1909]. Wilhelm Kübler, "The Scherl Monorail-road of Germany," *The American Machinist*, Vol. XXXII, Part 2, 1909, pp. 1135-1138; http://books.google.com.

[Landau 1960]. Lev Davidovich Landau and Evgeny Mikhailovich Lifshitz, *Mechanics*, Pergamon Press, Oxford, 1960.

[Lasche 1901]. Oskar Lasche, "High-Speed Electrical Traction in Germany," *Engineering*, Vol. 72, 1901, pp. 369–370, 558.

[Lohmeier 2005]. Dieter Lohmeier and Bernhardt Schell, *Einstein, Anschütz und der Kieler Kreiselkompass*, Raytheon Marine GmbH, Kiel, 2005.

[Lorenz 1904]. Hans Lorenz, "Die Wirkung eines Kreisels auf die Rollbewegung von Schiffen," *Physikalische Zeitschrift*, 5. Jahrgang, 1904, p. 27–32; http://books.google.com.

[Lucas 1918]. Alfred Lucas, *Des Phénomènes Gyroscopiques et de leurs Principales Applications à la Navigation*, Vol. IV, Challamel, Paris, 1918; http://hathitrust.org.

[MacMillan 1960]. William Duncan MacMillan, *Dynamics of Rigid Bodies*, Dover Publications, New York, 1960.

[Magnus 1954]. Wilhelm Magnus and Fritz Oberhettinger, *Formulas and Theorems for the Functions of Mathematical Physics*, Chelsea Publishing Co., New York, 1954.

[Malmström 1905]. Rurik Malmström, "Versuch einer Theorie der elektrolytischen Dissoziation unter Berücksichtigung der elektrischen Energie," Leipzig, 1905; http://books.google.com.

[Malmström 1909]. Rurik Malmström, "Die Theorie des Schlick'schen Schiffskreisels," *Acta societatis scientiarum Fennicæ*, Tom. XXXV, No. 2, 1909.

[Martienssen 1906]. Oscar Martienssen, "Die Verwendbarkeit des Rotationskompasses als Ersatz des magnetischen Kompasses," *Physikalische Zeitschrift*, Vol. 7, No. 15, 1906, pp. 535–543.

[Martienssen 1923]. Oscar Martienssen, "Die Entwicklung des Kreiselkompasses," *Zeitschrift des Vereines deutscher Ingenieure*, Vol. 67, No. 8, 1923, pp. 182–187.

[Maxwell 1870]. James Clerk Maxwell, "Address to the Mathematical and Physical Sections of the British Association," *Nature*, Vol. 2, No. 47, 1870, pp. 419–428; reprinted in *The Scientific Papers of James Clerk Maxwell*, Vol. II, Cambridge University Press, 1890, pp. 215–229; http://books.google.com.

[Meijaard 2007]. J. P. Meijaard, J. M. Papadopolous, A. Ruina, and A. L. Schwab, "Linearized dynamics equations for the balance and steer of a bicycle: a benchmark and review," *Proceedings of the Royal Society A*, Vol. 463, 2007, pp. 1955–1982; http://rspa.royalsocietypublishing.org.

[Meldau 1906]. Heinrich Meldau, "Nautik," *Encyklopädie der mathematischen Wissenschaften*, Band VI, Teil 1, Art. 5, B. G. Teubner, Leipzig, 1906–1925, pp. 297–372; http://gdz.sub.uni-goettingen.de.

[Mottelay 1910]. Paul F. Mottelay, "The Regnard Aeroplane," *Scientific American Supplement*, Vol. LXX, No. 1814, 1910, pp. 228–229; http://books.google.com.

[Noether 1985]. Gottfried Noether, "Fritz Noether (1884-194?)," *Integral Equations and Operator Theory*, Vol. 8, 1985, pp. 573–576.

[Parastaev 1990]. Andrei Parastaev, "Letter to the Editor," *Integral Equations and Operator Theory*, Vol. 13, 1990, pp. 303–305.

[Perry 1957]. John Perry, *Spinning Tops and Gyroscopic Motions*, Dover Publications, New York, 1957.

[Prandtl 1910]. Ludwig Prandtl, "Einige für die Flugtechnik wichtige Beziehungen aus der Mechanik," *Zeitschrift für Flugtechnik und Motorluftschiffahrt*, Jahrgang 1, 1910, pp. 3–6, 25–30, 61–64, 74–76; http://hathitrust.org.

[Rankine 1869]. William John Macquorn Rankine, "On the dynamical principles of the motion of velocipedes," *The Engineer*, Vol. 28, 1869, pp. 79, 129, 153, 175, Vol. 29, 1870, p. 2; http://gracesguide.co.uk.

[Regnard 1912]. Paul Louis Antoine Regnard, "Automatic Stabilizer for Aeroplanes and the Like," US Patent 1,015,837, 1912; http://uspto.-gov.

[Routh 1960]. Edward John Routh, *The Elementary Part of a Treatise on the Dynamics of a System of Rigid Bodies*, Dover Publications, New York, 1960.

[de Santis 1936]. Renato de Santis and Michele Russo, "Rolling of the S. S. 'Conte di Savoia' in Tank Experiments and at Sea," *Transactions of the Society of Naval Architects and Marine Engineers*, Vol. 44, 1936, pp. 169–194.

[Scherl 1909]. August Scherl, *Ein neues Schnellbahnsystem*, Druck und Verlag von August Scherl, 1909; http://commons.wikimedia.org.

[Schlick 1909]. Otto Schlick, "Der Schiffskreisel," *Jahrbuch der Schiffbautechnischen Gesellschaft*, Bd. 10, 1909, pp. 111–148.

[Schlote 1991]. Karl-Heinz Schlote, "Fritz Noether – Opfer zweier Diktaturen. Tod und Rehabilitierung," *NTM Shriftenreihe für Geschichte der Naturwissenschaften und Medizin*, Vol. 28, 1991, pp. 33–41.

[Schuler 1923]. Max Schuler, "Die Störung von Pendel- und Kreiselapparten durch Beschleunigung des Fahrzeuges," *Physikalische Zeitschrift*, Vol. 24, 1923, pp. 344–350.

[Sears 1898]. Walter J. Sears, "The Whitehead Torpedo," *Engineering*, Vol. 66, 1998, pp. 89–91; http://books.google.com.

[Siacci 1888]. Francesco Siacci, *Balistica*, F. Casanova, Torino, 1888; http://books.google.com.

[Silliman 1963]. Robert H. Silliman, "William Thomson: Smoke Rings and Nineteenth-Century Atomism," *Isis*, Vol. 54, No. 4, 1963, pp. 461–474; http://jstor.org.

[Sommerfeld 1904]. Arnold Sommerfeld to Felix Klein, 8 November 1904, in *Arnold Sommerfeld. Wissenschaftlicher Briefwechsel*, Band 1, 1892-1918, p. 238, edited by Michael Eckert and Karl Märker, GNT-Verlag and Deutsches Museum, 2000.

[Sommerfeld 1892]. Arnold Sommerfeld, "Mechanische Darstellung der electromagnetischen Erscheinungen in ruhenden Köpern," *Annalen der Physik*, Bd. 46, 1892, pp. 139–151; reprinted in Arnold Sommerfeld, *Gesammelte Schriften*, Bd. I, Friedr. Vieweg & Sohn, Braunschweig, 1968, pp. 404–416.

[Sommerfeld 1940]. Arnold Sommerfeld to the Russian Academy, 6 August 1940, carbon copy in Deutsches Museum Archive, NL 89, 024.

[Sommerfeld 1952]. Arnold Sommerfeld, *Mechanics*, Lectures on Theoretical Physics, Vol. I, translated from the fourth German edition by Martin O. Stern, Academic Press, 1952.

[Stäckel 1901]. Paul Gustav Stäckel, "Elementare Dynamik der Punktsysteme und starren Körper," *Encyklopädie der mathematischen Wissenschaften*, Band IV, Teil 1, Art. 6, B. G. Teubner, Leipzig, 1901–1908, pp. 435–684; http://gdz.sub.uni-goettingen.de.

[Stodola 1905]. Aurel Stodola, *Steam Turbines*, English translation by Louis C. Loewenstein, D. Van Nostrand, New York, 1905; http://books.-google.com.

[Thomson 1883]. Joseph John Thomson, *A Treatise on the Motion of Vortex Rings*, Macmillan and Co., London, 1883; http://books.-google.com.

[Thomson 1890]. William Thomson, *Mathematical and Physical Papers*, Vol. III, C. J. Clay and Sons, London, 1890; http://archive.org.

[Thomson 1910]. William Thomson, *Mathematical and Physical Papers*, Vol. IV, University Press, Cambridge, 1910, pp. 475–481; http://-books.google.com.

[Tomlinson 1980]. Norman Tomlinson, *Louis Brennan, Inventor Extraordinaire*, John Hallewell Publications, 1980.

[Tucker 1983–84]. David Gordon Tucker, "F. B. Behr's Development of the Lartigue Monorail: From Country Crawler to Electric Express," *Transactions of the Newcomen Society*, Vol. 55, 1983–84, pp. 131–152.

[Walker 1904]. Gilbert Thomas Walker, "Spiel und Sport," *Encyklopädie der mathematischen Wissenschaften*, Band IV, Teil 2, Art. 9, B. G. Teubner, Leipzig, 1904–1935, pp. 127–152, http://gdz.sub.uni-goettingen.de.

[Wechmann 1923]. Wilhelm Wechmann, "Gustav Wittfeld †," *Zentralblatt der Bauverwaltung*, 43. Jahrgang, Nr. 85/86, 1923, p. 515; http://kobv.de.

[Welz 1905]. Franz Welz (tr.), *The Berlin-Zossen Electric Railway Tests of 1903*, McGraw Publishing Co., New York, 1905; http://archive.org.

[Whipple 1899]. Francis John Welsh Whipple, "The stability of the motion of a bicycle," *Quarterly Journal of Pure and Applied Mathematics*, Vol. XXX, 1899, pp. 312–348; http://hruina.tam.cornell.edu.

[Whittaker 1964]. Edmund Taylor Whittaker, *Analytical Dynamics of Particle and Rigid Bodies*, Cambridge University Press, 1964; http:-//archive.org (1917 ed.).

[Whitney 1889]. William Dwight Whitney *The Century Dictionary*, The Century Co., New York, Vol. III, 1889; http://archive.org.

[Worthington 1920]. Arthur Mason Worthington, *Dynamics of Rotation*, Longmans, Green, and Co., London, 1920; http://books.-google.com.

Index

Printed in the United States
By Bookmasters